AF595338

Organic Optoelectronics: Photoelectronic Conversion Materials, Physics and Devices

Organic Optoelectronics: Photoelectronic Conversion Materials, Physics and Devices

Editors

Xinping Zhang
Baoquan Sun
Fujun Zhang

MDPI • Basel • Beijing • Wuhan • Barcelona • Belgrade • Manchester • Tokyo • Cluj • Tianjin

Editors
Xinping Zhang
Beijing University of Technology
China

Baoquan Sun
Soochow University
China

Fujun Zhang
Beijing Jiaotong University
China

Editorial Office
MDPI
St. Alban-Anlage 66
4052 Basel, Switzerland

This is a reprint of articles from the Special Issue published online in the open access journal *Crystals* (ISSN 2073-4352) (available at: https://www.mdpi.com/journal/crystals/special_issues/organic_optoelectronics).

For citation purposes, cite each article independently as indicated on the article page online and as indicated below:

LastName, A.A.; LastName, B.B.; LastName, C.C. Article Title. *Journal Name* **Year**, *Volume Number*, Page Range.

ISBN 978-3-0365-7672-5 (Hbk)
ISBN 978-3-0365-7673-2 (PDF)

Contents

Preface to "Organic Optoelectronics: Photoelectronic Conversion Materials, Physics and Devices"

Organic optoelectronics is a broad topic involving research interests in a wide range of fields. Physics, materials, structures, and devices based on molecular semiconductors can all be classified into this topic. Another classification may be made by considering light emitting, light harvesting, light conversion or detection processes, where examining how light interacts with semiconductor molecules is the main criterium. This Special Issue intends to make a collection of the most recent progresses in organic optoelectronics.

In order to focus the topics of this Special Issue on the most active fields, we have chosen a title covering photoelectronic conversion materials, physics, and devices. Research articles have been collected in relevance to the light-emitting performance of organic or hybrid semiconductors, photovoltaic or photodetection devices, micro-cavity and random lasers, and structural or morphological or crystallization properties of semiconductor molecules influencing the photoelectronic processes.

Xinping Zhang, Baoquan Sun , and Fujun Zhang
Editors

Article

A Theoretical Model of Quasicrystal Resonators: A Guided Optimization Approach

Libin Cui [1], Anwer Hayat [1], Linzheng Lv [2], Zhiyang Xu [1] and Tianrui Zhai [1,*]

[1] College of Physics and Optoelectronics, Faculty of Science, Beijing University of Technology, Beijing 100124, China; cuilb_68@bjut.edu.cn (L.C.); anwerhayatnoor@gmail.com (A.H.); xu.zhiyang@bjut.edu.cn (Z.X.)

[2] College of Continuing Education, Beijing University of Technology, Beijing 100124, China; lvlinzheng@bjut.edu.cn

* Correspondence: trzhai@bjut.edu.cn; Tel.: +86-010-6739-2184

Abstract: Fibonacci-spaced defect resonators were analytically investigated by cavity coupling, which exhibited a series of well-defined optical modes in fractals. The analytic model can be used to predict the output performance of microcavity lasers based on Fibonacci-spaced defect resonators, such as the mode number, resonant frequency, and Q factor. All results obtained by the analytical solution are in good consistency with that obtained by the finite-difference time-domain method. The simulation result shows that the Q factor of the resonant modes would increase dramatically with the appearance of narrower optical modes. The proposed theoretical model can be used to inversely design high performance polymer lasers based on the Fibonacci-spaced defect resonators.

Keywords: quasicrystal resonator; analytic model; cavity coupling

Citation: Cui, L.; Hayat, A.; Lv, L.; Xu, Z.; Zhai, T. A Theoretical Model of Quasicrystal Resonators: A Guided Optimization Approach. *Crystals* **2021**, *11*, 749. https://doi.org/10.3390/cryst11070749

Academic Editor: Valentina Domenici

Received: 12 May 2021
Accepted: 23 June 2021
Published: 26 June 2021

1. Introduction

The discovery of quasicrystals in condensed matter has revolutionized solid-state physics [1–3]. During the last few decades, photonic quasicrystals have been extensively discussed and studied [4–9]. Quasiperiodic structures are natural intermediate cases between periodicity and randomness, which provide more optical design possibilities and richness in the engineering performance of the optical devices. The design of optical devices based on the quasiperiodic structures can achieve a better performance than periodic ones for some specific optical applications, which opens new avenues in the quest for high performance optical devices [10–16].

One-dimensional (1D) optical quasicrystal lattices composed of multilayer stacks have two different dielectrics of permittivity, ε_1 and ε_2, arranged in deterministic generation rule, which exhibit long-range order but lack of translational symmetry. All these structures exhibit self-similar properties.

In the case of 1D optical quasicrystals the quasicrystalline long-range order results in a pseudogap similar to the bandgap of photonic crystals, while the lack of periodicity in the quasicrystal results in critically localized optical modes similar to the localization in the random systems; in other words, it can be considered as a defect effect. This leads to the appearance of allowed optical modes inside the forbidden 1D pseudogap. The optical modes exhibit fractal self-similar patterns in the transmission spectra that stem from the self-similarity of the underlying structure [17–19]. High Q-factor resonators can be attained due to the splitting and sharpness of the previous modes from the well-defined self-similar feature [20]. Moreover, the progressive fragmentation of the frequency spectrum gives rise to a series of optical modes, which means that in the limit of a large sample size their spectra become singular continuous, which provide many additional Bragg resonances for feedback, leading to a multicolor laser at arbitrarily chosen frequencies within the gain bandwidth [21,22].

The optical modes inside quasicrystals are confined in space but decay weakly rather than exponentially [23–25], which leads to the crosstalk between modes in the energy spectra and it is difficult to realize single-mode control. This is commonly referred to as the "non-resonant" case [4,5,26]. Resonant quasiperiodic multi-quantum-well stacks were proposed [27–29], which show a shorter emission lifetime and higher photoluminescence emission intensity than that in non-resonant conditions.

How are resonator properties such as wavelength, number of modes and output controlled by the quasiperiodicity? Many important concepts that are related to crystals, such as band structure, Bloch theorem, Brillouin zone, etc., are invalid in quasicrystals. The optical behavior of 1D optical quasicrystal lattices can be investigated numerically using the transfer matrix method [26,30–32] and the plane wave method [33]. However, the numerical solutions are not intuitive.

In this paper, we proposed a resonant Fibonacci-spaced multiple-defect-cavity structures, which exhibits properties of self-similar optical modes with a series of well-separated peaks. At the resonant condition, the system is resonant and strongly coupled. When there is an increase in the generation order, very narrow optical modes would appear and the Q factor of the resonant modes would increase exponentially with mode splitting. We developed a general theoretical model of the 1D quasicrystal resonator based on the coupled mode theory. The mode splitting, number and frequency of the resonant modes were explained qualitatively by an analytical solution. The theoretical model is completely consistent with the finite difference time domain (FDTD) simulation. We also demonstrate the inverse design of 1D Fibonacci-spaced resonators for the desired wavelength and amplitude of optical modes which can realize the broad color gamut laser display.

2. Theoretical Analysis

2.1. Theoretical Model of 1D Fibonacci-Spaced Defect Resonators

One dimensional resonant Fibonacci quasicrystals systems based on multiple defect cavities with two different interdefect distances satisfy the Fibonacci-chain rule between the long and short interdefect distances, as shown in Figure 1. The separations between the defect cavities are denoted by short(S) and long(L), respectively. The Fibonacci chain ($LSLLSLSL\ldots$) follows the construction rule that the next complete sequence is the present sequence plus the previous sequence, marking the first sequence as S and the second as L.

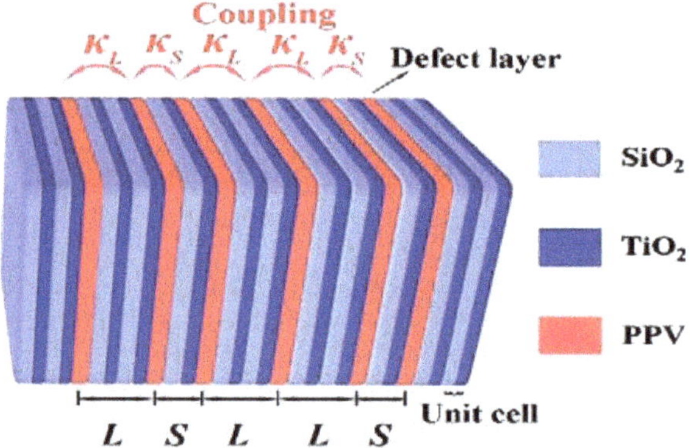

Figure 1. Scheme of 1D Fibonacci-spaced defect resonators. The PPV layers are inserted in the period stacks of SiO_2/TiO_2 structures, in which the PPV acts as both the defect cavity and the gain medium.

In the defect cavity, light is confined in a small volume, which enhances light–matter interactions and favors the laser behavior. The high-Q resonance and wavelength size of defect cavity make it extremely suitable for creating the coupled microcavity arrays. The widths of the interdefect cavity were determined from the resonant Bragg conditions, specifying the constructive interference of the waves reflected from the multiple defect cavities at the excitonic resonance.

The resonant electric field of each defect microcavity could enter into the other's field with a coupling constant κ. Similarly, the amount of the coupling can be defined by taking the overlap integral of the modes from the electric field as it enters into the other's field with a coupling strength. We have developed our theoretical model based on the coupled-wave theory. When the coupling effect between the cavities is ignored, there will be one resonant mode in the above system under the resonant condition. The field amplitude in the microcavity evolves over time as $exp(-i\omega t)$, therefore, the dynamic equations for resonance amplitudes can be written as: $\frac{da}{dt} = -i\omega a$, where a is the field amplitude in the microcavity, and ω is the resonant frequency [34]. Let us consider the cavity mode coupling effect between the cavities, as shown in Figure 1. We ignore the coupling between the non-adjacent cavities. According to the coupled-wave theory, the dynamic equations for the nth resonant cavity can be written in the following form [35]:

$$\frac{da_n}{dt} = -i\omega_n a_n + i\kappa_{n-1\ n} a_{n-1} + i\kappa_{n+1\ n} a_{n+1} \tag{1}$$

where a_n is the mode amplitude of the nth resonator, $\kappa_{n-1\ n}$ and $\kappa_{n+1\ n}$ present the coupling coefficients of the modes $n-1$ and $n+1$ coupled to the mode n, respectively. The Fourier transform of a_n is

$$a_n(t) = \int A_n(\omega) exp(-i\omega t) d\omega \tag{2}$$

Inserting the value of Equation (2) into Equation (1) and considering all N resonators, after solving, the coupling equations in the frequency domain for the given system can be expressed as:

$$\begin{bmatrix} \omega_1-\omega_0 & -\kappa_{21} & 0 & \cdots & 0 \\ -\kappa_{12} & \omega_2-\omega_0 & -\kappa_{32} & \cdots & 0 \\ \cdots & \cdots & \cdots & \cdots & \cdots \\ 0 & 0 & \cdots & \omega_{N-1}-\omega_0 & -\kappa_{NN-1} \\ 0 & 0 & \cdots & -\kappa_{N-1N} & \omega_N-\omega_0 \end{bmatrix} \begin{bmatrix} A_1 \\ A_2 \\ \cdots \\ A_{N-1} \\ A_N \end{bmatrix} = 0 \tag{3}$$

where $A_1, A_2 \cdots A_N$ are the complex amplitude of the N resonant cavities and ω_0 is the unperturbed Bragg frequency. Therefore, the above equations will lead a nonzero solutions when the determinant of the coefficient of equations D_N equals zero. Considering the coupling coefficient between the two cavities is irrelevant to the direction of coupling. In the two opposite directions, we suppose $\kappa_{ij} = \kappa_{ji} = \kappa_i$, D_N can be expressed as:

$$D_N = \begin{vmatrix} \omega_1-\omega_0 & -\kappa_1 & 0 & \cdots & 0 \\ -\kappa_1 & \omega_2-\omega_0 & -\kappa_2 & \cdots & 0 \\ \cdots & \cdots & \cdots & \cdots & \cdots \\ 0 & 0 & \cdots & \omega_{N-1}-\omega_0 & -\kappa_{N-1} \\ 0 & 0 & \cdots & -\kappa_{N-1} & \omega_N-\omega_0 \end{vmatrix} = 0 \tag{4}$$

According to Equation (4), we construct the coupling matrix as follows:

$$C_N = \begin{vmatrix} 0 & \kappa_1 & 0 & \cdots & 0 \\ \kappa_1 & 0 & \kappa_2 & \cdots & 0 \\ \cdots & \cdots & \cdots & \cdots & \cdots \\ 0 & 0 & \cdots & 0 & \kappa_{N-1} \\ 0 & 0 & \cdots & \kappa_{N-1} & 0 \end{vmatrix} \tag{5}$$

Equation (5) implies that diagonal elements are zero because the mode of the one cavity is not coupled to itself for the off diagonal elements $\kappa_{ij} = 0$, except $j = i \pm 1$, because coupling between the non-adjacent cavities should not be considered.

By comparing Equation (4) and Equation (5), we can obtain $D_N = |(\omega - \omega_0)E - C_N|$, where E is the unit matrix. Therefore, $\omega - \omega_0$ is the eigenvalue of C_N. As described in Equation (5), C_N is a real symmetric matrix, $(C_N)^T = C_N$. For N we can obtain the eigenvalue $\omega_1 - \omega_0$, $\omega_2 - \omega_0$ ... $\omega_N - \omega_0$ from the symmetry of C_N. It can be further proved by the matrix theory that these N eigenvalues are different. (The detailed derivation process is described in Appendix A). Thus, the cavity coupled modes split into the N resonance modes.

Based on the well-known matrix theory, we can obtain the resonance frequencies for the N defect microcavities arranged following the Fibonacci sequence as described in Figure 1. The coupling coefficients between the cavities with short and long separations are denoted by κ_S and κ_L, respectively. Therefore, Equation (4) can be written as:

$$D_N = \begin{vmatrix} \omega_1 - \omega_0 & -\kappa_L & 0 & \cdots & 0 \\ -\kappa_L & \omega_2 - \omega_0 & -\kappa_S & \cdots & 0 \\ \cdots & \cdots & \cdots & \cdots & \cdots \\ 0 & 0 & \cdots & \omega_{N-1} - \omega_0 & -\kappa_{N-1} \\ 0 & 0 & \cdots & -\kappa_{N-1} & \omega_N - \omega_0 \end{vmatrix} = 0 \quad (6)$$

After solving the linear equations, the resonant frequencies can be obtained. D_N can be expressed as the following recursive relation (See Appendix B for detailed derivation). Note that κ_{N-1} depends on the parity of N.

$$D_N = (\omega - \omega_0)D_{N-1} - \kappa_{N-1}^2 D_{N-2} \quad (7)$$

Once we know both D_2 and D_3, D_N can be extracted step by step using Equation (7).

2.2. Analytical Results and Discussion

We can figure out all the resultant resonant frequencies from the solution of Equation (6) and some results are described in Table 1.

Table 1. The frequencies of the coupled defect mode for $N = 1, 2, 3, 5$, respectively.

The Defect Numbers	The Frequencies of the Coupled Defect Modes
1	$\omega = \omega_0$
2	$\omega = \omega_0 \pm \Delta\omega = \omega_0 \pm \kappa_L$
3	$\omega = \omega_0, \omega = \omega_0 \pm \Delta\omega = \omega_0 \pm \sqrt{\kappa_S^2 + \kappa_L^2}$
5	$\omega = \omega_0, \omega = \omega_0 \pm \Delta\omega = \omega_0 \pm \kappa_L, \omega = \omega_0 \pm \Delta\omega = \omega_0 \pm \sqrt{\kappa_S^2 + 2\kappa_L^2}$

We noticed in the analytical solution that the resonant frequencies are distributed symmetrically about ω_0 and can be evaluated from the coupling coefficient κ_S and κ_L. If $\kappa_S = \kappa_L$, it is referred to a as periodic defect cavities chain, which has been studied in our previous work [36]. Similarly, the Fibonacci-spaced defect resonators exhibit unusual properties, which are very different from those of periodic and random systems, as we can see in Figure 2a, where we plot the evolution of resonant modes as generation order increase, which show self-similar properties of typical Cantor sets.

It could be proved by the matrix theory there exist a limit value for the maximum and minimum resonant frequencies in the above analytical solution. This means that in a limited spectral region the spectra are highly fragmented. For large generation order, the line width of the resonant modes decreases dramatically, which results in resonators with high Q-factors.

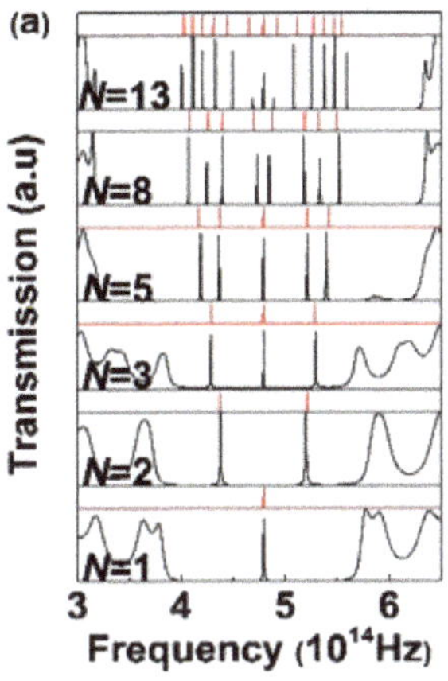

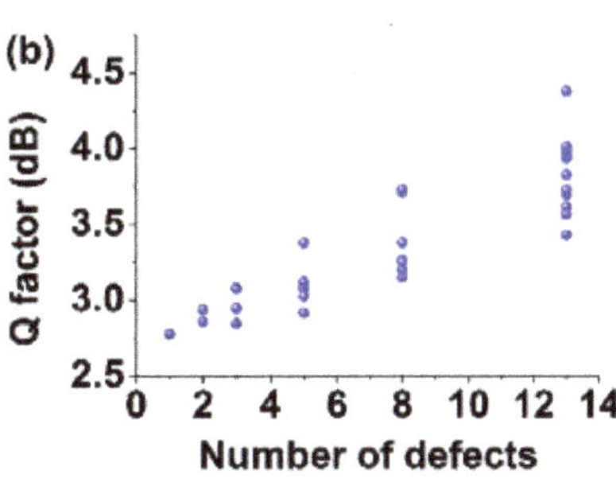

Figure 2. (**a**) Comparison of resonant frequencies of the Fibonacci-spaced defect resonators obtained from the theoretical model developed (the red line) and the simulation using FDTD (the black line) with different generation order for $\kappa_S = 0.42$, $\kappa_L = 0.27$. The results agree well, and the discrepancies are expected to vanish if the first and last coupling coefficients are modified due to external losses. (**b**) Q factor increases exponentially with mode splitting.

Due to their highly fragmented frequency spectra, a Fibonacci-spaced chain of defect cavities offers more resonant frequencies than periodic ones in a given frequency range for a given system length, which provides a higher degree of design and tuning flexibility. The analytical solution of the proposed theoretical model can be used to customize high performance microcavity resonators and it can be made via inverse design. The theoretical mode can also be extended to predict the characteristics of 2D quasiperiodic structures.

3. Simulation and Validation

To validate the theoretical analysis, we compare the theoretical results to the FDTD simulations. In the simulation, the refractive indices of SiO_2 and TiO_2 were chosen as 1.54 and 2.5, respectively, which were measured by a spectroscopic ellipsometer (ESNano, Ellitop). The thickness of the SiO_2/TiO_2 layer was 90/70 nm. The refractive index and the thickness of the MDMO-PPV (defect/gain layer) were about 1.67 and 94 nm, respectively. The important parameters were optimized by FDTD, so that the photonic band gap overlapped almost completely with the emission spectrum of MEMO-PPV, and the unperturbed defect mode frequency is exactly at the center of band-gap of the 1D photonic structure.

We found an excellent agreement in the resonant frequencies by comparing the results of the analytical model based on the coupling mode effect with FDTD simulation, as shown in Figure 2a. The deviations between the theory and the simulation are from the coupling of two cavities on both sides to the outside world. We expect this discrepancy vanish if the external losses are modified in the first and last coupling coefficients.

It can be observed from Figure 2a that every mode of any hierarchy will split into submodes to form the Cantor spectrum. In the third hierarchy there are three modes: the mode on the both sides will further split into submode for the global structure, and the middle mode will also split in the same manner for next stage. Thus, it turns out that the Fibonacci-spaced defect resonators have a perfect self-similar spectrum. The Q factor of the resonant modes would increase exponentially with the appearance of narrower optical modes, as shown in Figure 2b. Considering only the resonant modes in the center of the band, the Q value increases from 480 for $N = 1$ to 23,960 for $N = 13$, where the size of the device increases from 1.694 to 5.862 μm.

4. Inverse Design of Laser Resonators Based on the Fibonacci-Spaced Defect Resonators

Some features of the regular quasicrystals are not required for the operation of microcavity lasers. According to the analytical model, the regular quasicrystals can be modified to be an ideal resonator for microcavity lasers. For example, the mode splitting provides a simple, flexible and versatile approach for the design of high performance resonators [36–40]. It would allow us to engineer the behavior of the cavity mode simply by tailoring the separation and size of the defects. Here, the ratio $\alpha = \frac{L}{S}$ is arbitrary. For $L = S$ the structure becomes periodic and for τ equal to the golden mean 1.618, it becomes the canonical Fibonacci chain [41]. The simulation results show that $\Delta\omega$, which refers to frequency detuning from ω_0 due to coupling, decreases exponentially with increasing the ratio α, as shown in Figure 3.

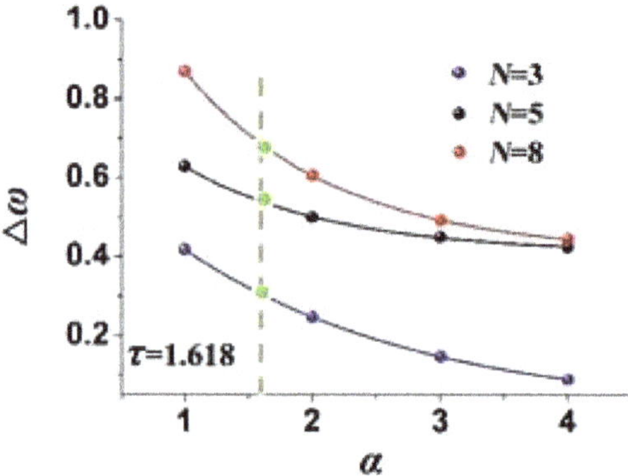

Figure 3. The relation between $\Delta\omega$ and α for $N = 3, 5, 8$. The dotted line indicates the positions on the curve at $\tau = 1.618$.

According the relation between $\Delta\omega$ and α and the analytical solutions mentioned above, we can optimize 1D quasicrystal resonators to obtain target optical modes at arbitrary positions in a broad spectral range.

In this section, we have studied inverse design of the 1D quasicrystals resonators to obtain desired laser emission, as shown in Figure 4. By adjusting the separation of the defects in above model, the optical modes can be fine-tuned across a broad-spectrum range, as shown in Figure 4a. Organic polymers have broad emission spectrum, which demonstrates excellent features in flexible spectrum modulation. It has been reported [42,43] that the intensity of the band-gap modes can be adjusted by controlling the phase shift of both reflecting facets. In Figure 4b, we have demonstrated the effect on the intensity of optical modes in the band-gap by changing the number of boundary layers. Based on the analytic model, the number of optical modes can be predicted, which corresponds to the number of defects, as shown in Figure 4c. The wavelength and intensity of the output emissions are calculated as points located in the CIE chromaticity diagram in Figure 4d. Thus, the CIE chromaticity demonstrates the output of the laser with broad color gamut.

Note that there is a slight difference between the resonant frequency of Fibonacci-spaced defect resonators obtained from the theoretical model and that obtained from FDTD, especially for high generation orders. It can be attributed to the fact that the coupling between non-adjacent cavities is not considered in the proposed method. Even so, the developed analytical model can be extended to a more general case of aperiodic plasmonic systems, which provides a simple and efficient route towards designing real systems with flexible, multispectral optical responses. It opens a new avenue in the quest for high performance optical devices.

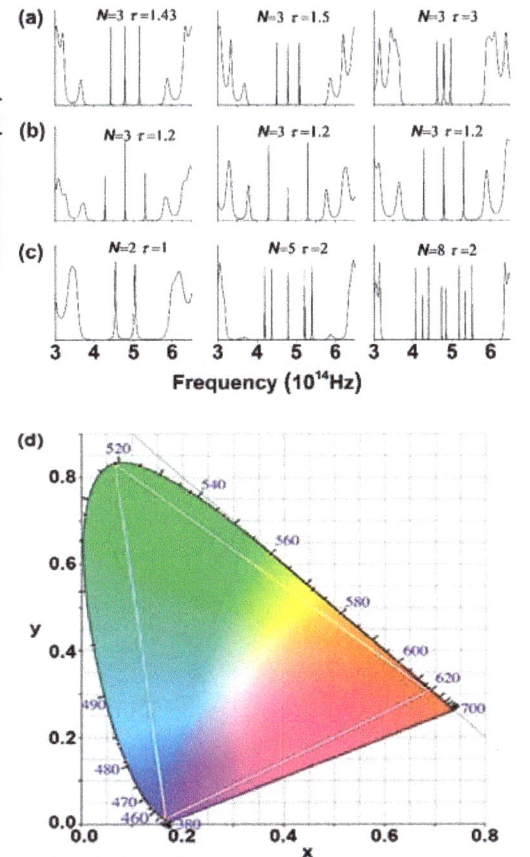

Figure 4. (**a**) Modes splitting for $N = 3$ with $\tau = 3$, 1.5, 1.43. (**b**) Intensity of optical modes for different boundary layers. (**c**) The number of optical modes for different number of defects. (**d**) Schematic diagram of the emission spectra for the R, G and B components.

5. Conclusions

In conclusion, we have presented 1D quasicrystal resonators which exhibit progressive fragmentation of the frequency spectrum as generation order increases and gives rise to quasicontinuous but well-defined optical modes in the limit of large sample size. The analytical solution of the proposed theoretical model provides precise control on single- or multifrequency across a broad spectral range, which can be used in the polymer laser-based display with broad color gamut and also can meet the operating requirements of high-resolution spectroscopy.

Author Contributions: Conceptualization, L.C. and T.Z.; Methodology, L.C. and L.L.; Validation, L.C. and T.Z.; Formal Analysis, L.C. and L.L.; Investigation, L.C. and Z.X.; Writing—Original Draft Preparation, L.C.; Writing—Review and Editing, L.C., A.H., and T.Z.; Supervision, T.Z.; Project Administration, T.Z.; Funding Acquisition, T.Z. All authors have read and agreed to the published version of the manuscript.

Funding: Beijing Natural Science Foundation (Z180015); National Natural Science Foundation of China (61822501).

Institutional Review Board Statement: Not applicable.

Informed Consent Statement: Not applicable.

Data Availability Statement: Data sharing not available.

Conflicts of Interest: The authors declare no conflict of interest.

Appendix A

In the first step, keeping in view the concept of matrix theory, if C_N is a real symmetric matrix, it can be deduced that C_N has N real eigenvalues. In the second step, if we have $D_N = (\omega - \omega_0)E - C_N = (d_{ij})$, then the cofactor of d_{N1} can be arranged in the following form:

$$\begin{vmatrix} -\kappa_1 & 0 & \cdots & 0 & 0 \\ \omega-\omega_0 & -\kappa_2 & \cdots & 0 & 0 \\ \cdots & \cdots & \cdots & \cdots & \cdots \\ 0 & 0 & \cdots & -\kappa_{N-2} & 0 \\ 0 & 0 & \cdots & \omega-\omega_0 & -\kappa_{N-1} \end{vmatrix}_{N-1} = (-1)^{N-1}\kappa_1\kappa_2\cdots\kappa_{N-1} \neq 0 \quad \text{(A1)}$$

Thus, the rank of D_N is greater than or equal to 1, that is $R(D_N) \geq N-1$. On the other hand, $R(D_N) \leq N-1$ because $\omega - \omega_0$ is the eigenvalue of C_N. Therefore, we can obtain $R(D_N) = N-1$ for any eigenvalue of C_N. From $R(D_N) = N-1$, we can derive that N real eigenvalues of C_N are different. In summary, it can be concluded that C_N has N different real eigenvalues.

Appendix B

D_N is the determinant of tridiagonal matrix in Equation (6). It can be written as:

$$D_N = \begin{vmatrix} \omega-\omega_0 & -\kappa_L & 0 & \cdots & 0 \\ -\kappa_L & \omega-\omega_0 & -\kappa_S & \cdots & 0 \\ \cdots & \cdots & \cdots & \cdots & \cdots \\ 0 & 0 & \cdots & \omega-\omega_0 & -\kappa_{N-1} \\ 0 & 0 & \cdots & -\kappa_{N-1} & \omega-\omega_0 \end{vmatrix} =$$

$$(\omega-\omega_0)\begin{vmatrix} \omega-\omega_0 & -\kappa_L & 0 & \cdots & 0 \\ -\kappa_L & \omega-\omega_0 & -\kappa_S & \cdots & 0 \\ \cdots & \cdots & \cdots & \cdots & \cdots \\ 0 & 0 & \cdots & \omega-\omega_0 & -\kappa_{N-2} \\ 0 & 0 & \cdots & -\kappa_{N-2} & \omega-\omega_0 \end{vmatrix}_{N-1} \quad \text{(A2)}$$

$$-\kappa_{N-1}^2\begin{vmatrix} \omega-\omega_0 & -\kappa_L & 0 & \cdots & 0 \\ -\kappa_L & \omega-\omega_0 & -\kappa_S & \cdots & 0 \\ \cdots & \cdots & \cdots & \cdots & \cdots \\ 0 & 0 & \cdots & \omega-\omega_0 & -\kappa_{N-3} \\ 0 & 0 & \cdots & -\kappa_{N-3} & \omega-\omega_0 \end{vmatrix}_{N-2}$$

Thus, from Equation (A2), D_N can be expressed as the following recursive relation:

$$D_N = (\omega-\omega_0)D_{N-1} - \kappa_{N-1}^2 D_{N-2}$$

where D_{N-1} and D_{N-2} are the determinant of the bottom-right $(N-1) \times (N-1)$ and $(N-2) \times (N-2)$ submatrix of D_N respectively.

References

1. Shechtman, D.; Blech, I.; Gratias, D.; Cahn, J.W. Metallic phase with long-range orientational order and no translational symmetry. *Phys. Rev. Lett.* **1984**, *53*, 1951–1953. [CrossRef]
2. Kraus, Y.E.; Aahini, Y.; Ringel, Z.; Verbin, M.; Zilberberg, O. Topological states and adiabatic pumping in quasicrystals. *Phys. Rev. Lett.* **2012**, *109*, 106402. [CrossRef] [PubMed]

3. Bourne, C.; Prodan, E. Non-commutative chern numbers for generic aperiodic discrete systems. *J. Phys. A Math. Theor.* **2018**, *51*, 235202. [CrossRef]
4. Kohmoto, M.; Sutherland, B.; Iguchi, K. Localization of optics: Quasiperiodic media. *Phys. Rev. Lett.* **1987**, *58*, 2436–2438. [CrossRef] [PubMed]
5. Nguyen, T.D.; Nahata, A.; Vardeny, Z.V. Measurement of surface plasmon correlation length differences using Fibonacci deterministic hole arrays. *Opt. Express* **2012**, *20*, 15222–15231. [CrossRef]
6. Vitiello, M.S.; Nobile, M.; Ronzani, A.; Tredicucci, A.; Castellano, F.; Talora, V.; Li, L.; Linfield, E.H.; Davies, A.G. Photonic quasi-crystal terahertz lasers. *Nat. Commun.* **2014**, *5*, 5884. [CrossRef]
7. Boguslawski, M.; Lučić, N.M.; Diebel, F.; Timotijević, D.V.; Denz, C.; Jović Savić, D.M. Light localization in optically induced deterministic aperiodic Fibonacci lattices. *Optica* **2016**, *3*, 711–717. [CrossRef]
8. Schokker, A.H.; Koenderink, A.F. Lasing in quasi-periodic and aperiodic Plasmon lattices. *Optica* **2016**, *3*, 686–693. [CrossRef]
9. Negro, L.D.; Chen, Y.; Sgrignuoli, F. Aperiodic photonics of elliptic curves. *Crystals* **2019**, *9*, 482. [CrossRef]
10. Moretti, L.; Rea, I.; Stefano, L.D.; Rendina, I. Periodic versus aperiodic: Enhancing the sensitivity of porous silicon based optical sensors. *Appl. Phys. Lett.* **2007**, *90*, 191112. [CrossRef]
11. Makarava, L.N.; Nazarov, M.M.; Ozheredov, I.A.; Shkurinov, A.P.; Smirnov, A.G.; Zhukovsky, S.V. Fibonacci-like photonic structure for femtosecond pulse compression. *Phys. Rev. E* **2007**, *75*, 036609. [CrossRef] [PubMed]
12. Macia, E. Exploiting aperiodic designs in nanophotonic devices. *Rep. Prog. Phys.* **2012**, *75*, 036502. [CrossRef]
13. Negro, L.D.; Boriskina, S.V. Deterministic aperiodic nanostructures for photonics and plasmonics applications. *Laser Photonics Rev.* **2012**, *6*, 178–218. [CrossRef]
14. Barriuso, A.G.; Monzon, J.J.; Yonte, T.; Felipe, A.; Soto, L. Omnidirectional reflection from generalized Fibonacci quasicrystals. *Opt. Express* **2013**, *21*, 30039–30053. [CrossRef]
15. Biasco, S.; Li, L.; Linfield, E.; Davies, A.; Vitiello, M. Multimode, aperiodic terahertz surface-emitting laser resonators. *Photonics* **2016**, *3*, 32. [CrossRef]
16. Davis, M.S.; Zhu, W.; Xu, T.; Lee, J.K.; Lezec, H.J.; Agrawal, A. Aperiodic nanoplasmonic devices for directional colour filtering and sensing. *Nat. Commun.* **2017**, *8*, 1347. [CrossRef]
17. Vasconcelos, M.S.; Albuquerque, E.L. Transmission fingerprints in quasiperiodic dielectric multilayers. *Phys. Rev. B* **1999**, *59*, 11128–11131. [CrossRef]
18. Thiem, S.; Schreiber, M. Local symmetry dynamics in one-dimensional aperiodic lattices: A numerical study. *Nonlinear Dynam.* **2014**, *78*, 71–91.
19. Tanese, D.; Gurevich, E.; Baboux, F.; Jacqmin, T.; Lema^itre, A.; Galopin, I.S.E.; Amo, A.; Bloch, J.; Akkermans, E. Fractal energy spectrum of a polariton gas in a Fibonacci quasi-periodic potential. *Phys. Rev. Lett.* **2014**, *112*, 146404–146409. [CrossRef]
20. Karman, G.P.; Mcdonald, G.S.; New, G.H.C.; Woerdman, J.P. Fractal modes in unstable resonators. *Nature* **1999**, *402*, 138. [CrossRef]
21. Mahler, L.; Tredicucci, A.; Beltram, F.; Walther, C.; Faist, J.; Beere, H.E.; Ritchie, D.A.; Wiersma, D.S. Quasi-periodic distributed feedback laser. *Nat. Photon.* **2010**, *4*, 165–169. [CrossRef]
22. Biasco, S.; Ciavatti, A.; Li, L.; Davies, A.G.; Linfield, E.H.; Beere, H.; Ritchie, D.; Vitiello, M.S. Highly efficient surface-emitting semiconductor lasers exploiting quasi-crystalline distributed feedback photonic patterns. *Light Sci. Appl.* **2020**, *54*, 1–11. [CrossRef]
23. Maciá, E. Physical nature of critical modes in Fibonacci quasicrystals. *Phys. Rev. B* **1999**, *60*, 10032–10036. [CrossRef]
24. Negro, L.D.; Oton, C.J.; Gaburro, Z.; Pavesi, L.; Johnson, P.; Lagendijk, A.; Righini, R.; Colocci, M.; Wiersma, D.S. Light transport through the band-edge states of Fibonacci quasicrystals. *Phys. Rev. Lett.* **2003**, *90*, 055501. [CrossRef]
25. Ghulinyan, M.; Oton, C.J.; Negro, L.D.; Pavesi, L.; Sapienza, R.; Colocci, M.; Wiersma, D.S. Light-pulse propagation in Fibonacci quasicrystals. *Phys. Rev. B* **2005**, *71*, 094204. [CrossRef]
26. Gellermann, W.; Kohmoto, M.; Sutherland, B.; Taylor, P.C. Localization of light waves in Fibonacci dielectric multilayers. *Phys. Rev. Lett.* **1994**, *72*, 633–636. [CrossRef]
27. Hendrickson, J.; Richards, B.C.; Sweet, J.; Khitrova, G.; Poddubny, A.N.; Ivchenko, E.L.; Wegener, M.; Gibbs, H.M. Excitonic polaritons in Fibonacci quasicrystals. *Opt. Express* **2008**, *16*, 15382–15387. [CrossRef]
28. Hsueh, W.J.; Chang, C.H.; Lin, C.T. Exciton photoluminescence in resonant quasi-periodic thue–morse quantum wells. *Opt. Lett.* **2014**, *39*, 489–492. [CrossRef] [PubMed]
29. Chang, C.H.; Chen, C.H.; Tsao, C.W.; Hsueh, W.J. Superradiant modes in resonant quasi-periodic double-period quantum wells. *Opt. Express* **2015**, *23*, 11946–11951. [CrossRef] [PubMed]
30. Kohmoto, M.; Kadanoff, L.P.; Tang, C. Localization problem in one dimension: Mapping and escape. *Phys. Rev. Lett.* **1983**, *50*, 1870–1872. [CrossRef]
31. Wang, X.; Grimm, U.; Schreiber, M. Trace and antitrace maps for aperiodic sequences: Extensions and applications. *Phys. Rev. B* **2000**, *62*, 14020–14031. [CrossRef]
32. Zhang, H.F.; Liu, S.; Kong, X.K.; Bian, B.R.; Zhao, X. Properties of omnidirectional photonic band gaps in Fibonacci quasi-periodic one-dimensional superconductor photonic crystals. *Prog. Electromagn. Res. B* **2012**, *40*, 415–431. [CrossRef]
33. Rychły, J.; Mieszczak, S.; Kłos, J.W. Spin waves in planar quasicrystal of Penrose tiling. *J. Magn. Magn. Mater.* **2018**, *450*, 18–23. [CrossRef]
34. Huang, W.P. Coupled-mode theory for optical waveguides: An overview. *JOSA A* **1994**, *11*, 963–983. [CrossRef]

35. Hardy, A.; Streifer, W. Coupled mode theory of parallel waveguides. *J. Lightwave Technol.* **1985**, *3*, 1135–1146. [CrossRef]
36. Cui, L.; Zhang, S.; Lv, L.; Xu, Z.; Hayat, A.; Zhai, T. Effects of cavity coupling on 1D defect modes: A theoretical model. *OSA Contin.* **2020**, *3*, 1408–1416. [CrossRef]
37. Zhang, S.; Tong, J.; Chen, C.; Cao, F.; Liang, C.; Song, Y.; Zhai, T.; Zhang, X. Controlling the performance of polymer lasers via the cavity coupling. *Polymers* **2019**, *11*, 764. [CrossRef] [PubMed]
38. Zhang, S.; Cui, L.; Zhang, X.; Tong, J.; Zhai, T. Tunable polymer lasing in chirped cavities. *Opt. Express* **2020**, *28*, 2809–2817. [CrossRef] [PubMed]
39. Hayat, A.; Tong, J.; Chen, C.; Niu, L.; Aziz, G.; Zhai, T.; Zhang, X. Multi-wavelengh colloidal quantum dot laser in distributed feedback cavities. *Sci. China Inf. Sci.* **2020**, *63*, 1–7. [CrossRef]
40. Wong, Y.; Jia, H.; Jian, A.; Lei, D.; Abed, A.I.E.; Zhang, X. Enhancing plasmonic hot-carrier generation by strong coupling of multiple resonant modes. *Nanoscale* **2021**, *13*, 2731–3310. [CrossRef]
41. Werchner, M.; Schafer, M.; Kira, M.; Koch, S.W.; Sweet, J.; Olitzky, J.D.; Hendrickson, J.; Richards, B.C.; Khitrova, G.; Gibbs, H.M.; et al. One dimensional resonant Fibonacci quasicrystals: Noncanonical linear and canonical nonlinear effects. *Opt. Express* **2009**, *17*, 6813–6828. [CrossRef]
42. Whiteaway, J.E.A.; Garrett, B.; Thompson, G.H.B.; Collar, A.J.; Armistead, C.J.; Fice, M.J. The static and dynamic characteristics of single and multiple phase-shifted DFB laser structures. *IEEE J. Quantum Electron.* **1992**, *28*, 1277–1293. [CrossRef]
43. Zhou, Y.; Shi, Y.; Chen, X.; Li, S.; Li, J. Numerical study of an asymmetric equivalent $\frac{\lambda}{4}$ phase shift semiconductor laser for use in laser arrays. *IEEE J. Quantum Electron.* **2011**, *47*, 534–540. [CrossRef]

MDPI

Article

Vortex Laser Based on a Plasmonic Ring Cavity

Xingyuan Wang [1,*], Xiaoyong Hu [2,*] and Tianrui Zhai [3,*]

1 College of Mathematics and Physics, Beijing University of Chemical Technology, Beijing 100029, China
2 State Key Laboratory for Mesoscopic Physics, Department of Physics, Collaborative Innovation Center of Quantum Matter, Peking University, Beijing 100871, China
3 School of Physics and Optoelectronics, Faculty of Science, Beijing University of Technology, Beijing 100124, China
* Correspondence: wang_xingyuan@mail.buct.edu.cn (X.W.); xiaoyonghu@pku.edu.cn (X.H.); trzhai@bjut.edu.cn (T.Z.)

Abstract: The orbital angular momentum (OAM) of the structure light is viewed as a candidate for enhancing the capacity of information processing. Microring has advantages in realizing the compact lasers required for on-chip applications. However, as the clockwise and counterclockwise whispering gallery modes (WGM) appear simultaneously, the emitted light from the normal microring does not possess net OAM. Here, we propose an OAM laser based on the standing-wave WGMs containing clockwise and counterclockwise WGM components. Due to the inhomogeneous intensity distribution of the standing-wave WGM, the single-mode lasing for the OAM light can be realized. Besides, the OAM of the emitted light can be designed on demand. The principle and properties of the proposed laser are demonstrated by numerical simulations. This work paves the way for exploring a single-mode OAM laser based on the plasmonic standing-wave WGMs at the microscale, which can be served as a basic building block for on-chip optical devices.

Keywords: micro-nano photonics devices; optical microcavity; micro-nano vortex laser; orbital angular momentum; plasmonic devices

Citation: Wang, X.; Hu, X.; Zhai, T. Vortex Laser Based on a Plasmonic Ring Cavity. *Crystals* **2021**, *11*, 901. https://doi.org/10.3390/cryst11080901

Academic Editor: Anna Paola Caricato

Received: 19 June 2021
Accepted: 29 July 2021
Published: 31 July 2021

Publisher's Note: MDPI stays neutral with regard to jurisdictional claims in published maps and institutional affiliations.

1. Introduction

After the knowledge about the linear momentum and spin angular momentum of light, a breakthrough is the recognition of the orbital angular momentum (OAM) of light [1]. The OAM is a completely new degree of freedom. In theory, the OAM light provides infinite orthogonal modes. As a result, the OAM states can greatly expand the information capacity, which has directed attention to applications for optical communication [2] and quantum information [3]. Since the first demonstration of a laser in 1960 [4], various lasers have been investigated, such as the plasmonic nanolaser [5], the random laser [6–8], the photonic crystal nanocavity laser [9], and the exciton-polariton laser [10], and so forth. Advances in nano-technology make it possible to manipulate the light on a chip [11–15]. Because the generation of micro-scale OAM light is of great significance to on-chip applications, it has received extensive attention.

Optical vortex beams with helical phase front carry OAM. In order to obtain OAM light, one can modulate the wave front of a coherent light. The wave-front modulations are mostly based on changing the optical path difference, employing a geometry phase, or controlling diffraction on demand with suitable optical elements, such as spiral phase plates, lens, Pancharatnam–Berry phase elements, and holograph [16–23] and so forth. In addition, the direct generation of OAM light within the laser cavity has been developed, to the great attention of researchers [24–34]. Microring has advantages in realizing compact lasers. However, the conventional microring cavity supports the standing-wave WGM with the same weight of clockwise (CW) and counterclockwise (CCW) components, resulting in a net zero OAM of the emitted light. In order to obtain emitted light with OAM, much attention is paid to breaking the rotation symmetry between the two counter-propagating

WGMs in the ring cavity. For instance, the OAM laser can be made based on the traveling-wave WGM at the non-Hermitian exceptional point of a ring cavity [34]. Despite these impressive advances, designing controllable single-mode OAM lasers at the microscale are still challenging.

Here, we propose a vector OAM laser based on the plasmonic standing-wave WGMs. In this framework, the OAM light is constructed by modulating the scattered waves from the standing-wave WGM in the plasmonic ring cavity, and the desired single-mode lasing can be achieved.

2. Principle and Results

Figure 1a shows the schematic of an OAM laser. The ring cavity is a coaxial cylindrical structure with a silver–InGaAsP–silver geometry. The bottom of the cavity is encapsulated by silver, so only the upper face of the ring is reserved as the port of the OAM-light emission. This microcavity supports plasmonic mode, for which the electric field component with polarization along the radial direction in the cavity plane (x-y plane) is dominant. We construct the OAM light based on the standing-wave WGMs. For the standing wave, the oscillations between two adjacent nodes are in phase, while the oscillations on the opposite sides of a node are in anti-phase, that is, there is a π phase jump at each node. However, the OAM light beam possesses helical phase fronts. In order to obtain OAM light, we introduce tiny effective-refractive-index perturbations at discrete positions along the azimuthal direction of the InGaAsP ring (the effective-refractive-index perturbations are in the region with $r_{\text{inner}} \le r \le r_{\text{outer}}$ and $0 \le z \le h_{\text{InGaAsP}}$, where r_{inner}, r_{outer} and h_{InGaAsP} are the inner radius, outer radius, and height of the InGaAsP ring, respectively). The tiny effective-refractive-index perturbations will couple the standing-wave WGM to the OAM mode by extracting the phase of the standing wave and shifting it to the phase of the OAM mode. In detail, we introduce $2m$ effective-refractive-index perturbations at discrete parts in the ring. Each individual section of the effective-refractive-index perturbations has an angular width $\Delta\theta = \pi/2m$ along the azimuthal direction of the ring. The azimuthal coordinates at the centrals of these $2m$ sections are $\theta_1(n) = \pi n/m$ $(n = 1, 2, 3, \ldots, 2m)$, respectively. For convenience, we define these $2m$ sections of the ring as region 1, and define other parts of the ring as region 2 (region 2 of the ring also consists of $2m$ sections, where each individual section has an angular width $\Delta\theta$, and the azimuthal coordinates at the centrals of these sections are $\theta_2(n) = \pi n/m + \Delta\theta$ $(n = 1, 2, 3, \ldots, 2m)$, respectively). In this framework, the standing-wave WGM with azimuthal mode number $l = m$ can be coupled to the desired OAM light. The effective refractive indexes in the ring ($r_{\text{inner}} \le r \le r_{\text{outer}}$, $0 \le z \le h_{\text{InGaAsP}}$) along the azimuthal direction (θ) are

$$n_G = \begin{cases} n_{\text{InGaAsP}} + \delta + \delta i + \delta e^{i\phi_n}, & \theta_1(n) - \frac{\Delta\theta}{2} \le \theta \le \theta_1(n) + \frac{\Delta\theta}{2} (n = 1,3,5\ldots 2m-1) \\ n_{\text{InGaAsP}} + \delta + \delta i + \delta e^{i(\phi_n+\pi)}, & \theta_1(n) - \frac{\Delta\theta}{2} \le \theta \le \theta_1(n) + \frac{\Delta\theta}{2} (n = 2,4,6\ldots 2m) \\ n_{\text{InGaAsP}} + n'' i, & \theta_2(n) - \frac{\Delta\theta}{2} \le \theta \le \theta_2(n) + \frac{\Delta\theta}{2} (n = 1,2,3\ldots 2m) \end{cases} \quad (1)$$

where the tiny perturbations $\delta e^{i\phi_n}$ and $\delta e^{i(\phi_n+\pi)}$ will couple the standing-wave WGMs with $l = m$ into the desired OAM light. Because the oscillations of the standing-wave WGMs jump a π phase at each amplitude node, a π is needed in the term $\delta e^{i(\phi_n+\pi)}$. $\phi_n = qn\pi/m$ $(n = 1, 2, 3, \ldots, 2m)$. The OAM carried by the emitted light is related to q $(q = 0, \pm1, \pm2, \ldots)$. The ϕ_n and $\phi_n + \pi$ can be controlled by tuning the real and imaginary part of the perturbation terms. Researchers have explored some methods to modulate the effective dielectric constant or effective refractive index [34,35]. The loss $n''i$ can be used to suppress the lasing of the competing modes. It is worth noting that the $\delta + \delta i$ is not required. However, in order to facilitate the practical effective refractive index modulations, $\delta + \delta i$ is used in our theoretical model. In this case, $\delta + \delta i + \delta e^{i\phi_n}$ and $\delta + \delta i + \delta e^{i(\phi_n+\pi)}$ $(\delta > 0)$ can be achieved without resorting to gain and negative real parts. In this paper, we adopt the following definition: the material corresponds to loss material when the imaginary part of the refractive index of the material is positive.

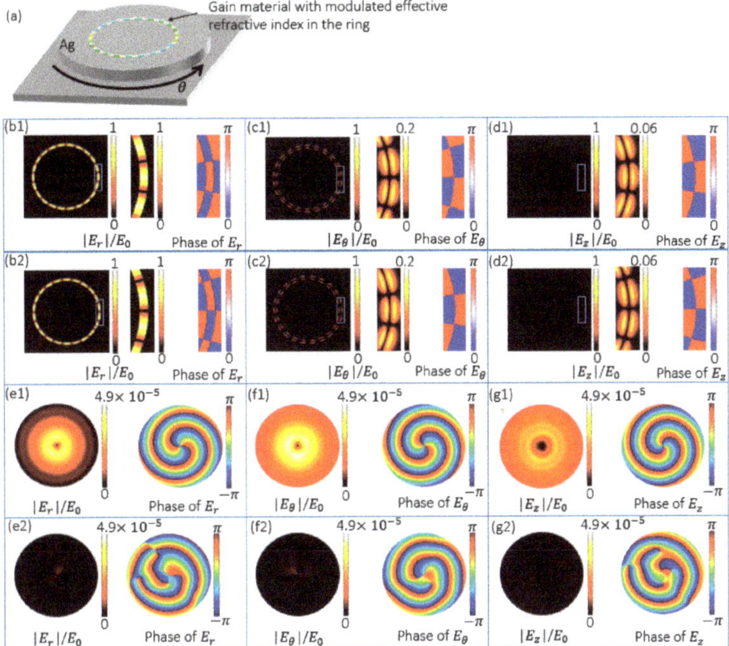

Figure 1. The cavity mode and the emitted mode of the OAM laser for $l = 11$. (**a**) Schematic of an OAM laser. The $z = 0$ is at the bottom of the InGaAsP ring, and the $(x = 0, y = 0)$ is at the center of the ring. (**b1**)–(**g1**), mode 1 inside the cavity [(**b1**)–(**d1**)] and the corresponding emission mode [(**e1**)–(**g1**)]. (**b2**)–(**g2**). Mode 2 inside the cavity [(**b2**)–(**d2**)] and the corresponding emission mode [(**e2**)–(**g2**)]. (**b1**),(**b2**), the intensity distributions $|E_r|/E_0$ of the radial component of the electric field inside the cavity (the $|E_r|/E_0$ and the phase of E_r in the area marked by rectangular box is also presented in the enlarged figures). (**c1**),(**c2**), the intensity distributions $|E_\theta|/E_0$ of the azimuthal component of the electric field inside the cavity (the $|E_\theta|/E_0$ and the phase of E_θ in the area marked by rectangular box is also presented in the enlarged figures). (**d1**),(**d2**), the intensity distributions $|E_z|/E_0$ of the z -component of the electric field inside the cavity (the $|E_z|/E_0$ and the phase of E_z in the area marked by rectangular box is also presented in the enlarged figures). (**b1**)–(**d1**), (**b2**)–(**d2**) are the mode properties at $z = h_{\text{InGaAsP}}/2$. (**e1**)–(**g1**), (**e2**)–(**g2**) are in the cross section (the radius of the area is 5000 nm) of the emission modes at 2 μm above the upper surface of the cavity. E_0 is the maximum of the $|E_r|$ in the (**b1**). The green lines in the enlarged figures are the boundaries of different sections. The $q = 2$ is used. The eigenfrequencies of the two modes are $1.9999 \times 10^{14} + i6.25 \times 10^{11}$ Hz and $2.0031 \times 10^{14} + i9.45 \times 10^{11}$ Hz, respectively. These results are calculated by the Comsol Multiphysics.

Now, we carry out 3D full wave simulations (via the commercial software COMSOL Multiphysics) to demonstrate the OAM laser. We take a standing-wave WGM with $l = 11$ in the microcavity as an example. The effective-refractive-index perturbations in region 1 consist of $2m$ discrete sections with an angular width $\Delta\theta = \pi/2m$, where $m = 11$. The azimuthal coordinate at the center of each section of region 1 is $\theta_1(n) = n\pi/11$, where $n = 1, 2, 3, \ldots, 22$. Then, the effective refractive indexes in the ring are

$$n_G = \begin{cases} n_{\text{InGaAsP}} + 0.006 + 0.006i + 0.006e^{i\phi_n}, & \frac{n\pi}{11} - \frac{\Delta\theta}{2} \le \theta \le \frac{n\pi}{11} + \frac{\Delta\theta}{2} (n = 1,3,5\ldots21) \\ n_{\text{InGaAsP}} + 0.006 + 0.006i + 0.006e^{i(\phi_n+\pi)}, & \frac{n\pi}{11} - \frac{\Delta\theta}{2} \le \theta \le \frac{n\pi}{11} + \frac{\Delta\theta}{2} (n = 2,4,6\ldots22) \\ n_{\text{InGaAsP}} + n''i, & \frac{n\pi}{11} + \frac{\Delta\theta}{2} \le \theta \le \frac{n\pi}{11} + \frac{3\Delta\theta}{2} (n = 1,2,3\ldots22) \end{cases} \quad (2)$$

where $\phi_n = qn\pi/11$, $n_{\text{InGaAsP}} = 3.34$, and $n'' = 0.012$. In the simulations, we consider the condition that the silver is at a low temperature, and use the refractive index

$n_{\text{Ag}} = 0.0014 + 10.9741i$ at 4.5 K [5], which has low metal loss. In this paper, we adopt the following definition: the mode corresponds to loss mode when the imaginary part of the eigenfrequency of the mode is positive. The inner radius, width and height of the InGaAsP ring are 620 nm, 50 nm and 200 nm, respectively.

Figure 1 presents the details of the standing waves in the cavity and the emitted modes for $l = 11$. Here, the laser cavity with $q = 2$ is taken as an example, and the z-direction mode number and radial mode number of the cavity mode are fixed. The $|E_r|$, $|E_z|$, and $|E_\theta|$ show obvious nodes along the azimuthal direction of the ring [Figure 1(b1)–(d1),(b2)–(d2)], which is the typical character of the standing wave modes. Besides, the standing-wave nature of the cavity modes is further confirmed by the phases of E_r, E_z, and E_θ of the cavity modes (the last figures of the Figure 1(b1)–(d1),(b2)–(d2)). Because the cavity mode is coupled into the emitted OAM light by the tiny perturbations in region 1, mode 1 and mode 2 with different distributions in the cavity exhibit distinct emission characteristics. For mode 1, the azimuthal antinodes of E_r (the dominant component of the electric field) match the sections of region 1 (Figure 1(b1),(d1)). Besides, the phase of the E_r of mode 1 in each individual section of region 1 is the same, and the E_r of mode 1 has a phase difference π between adjacent sections of region 1 (Figure 1(b1)). Thus, mode 1 is coupled to the desired OAM mode as shown by Figure 1(e1)–(g1), in which the E_r, E_z and E_θ all show phase characteristics of the OAM light. For mode 2, only the azimuthal antinodes of E_θ match the sections of region 1 (Figure 1(c2)), while the azimuthal antinodes of E_r and E_z match the sections of region 2 [Figure 1(b2),(d2)]. Besides, the E_r, E_θ, and E_z of mode 2 in each individual section of region 1 all simultaneously contain mode oscillations with opposite phases. Thus, the emission from mode 2 to the OAM mode is suppressed as shown in Figure 1(e2)–(g2) (the intensity of mode 1 and mode 2 in the cavity is nearly the same. However, compared with the bright color of the emitted light from mode 1 (Figure 1(e1)–(g1)), the emitted light from mode 2 (Figure 1(e2)–(g2)) shows a nearly dark color, indicating the suppression of the emitted light from mode 2.) This property will help suppress unwanted components in the emitted light.

Figure 2 shows that the OAM of the emitted light can be tuned by tuning the tiny modulation of the effective refractive index (tuning q in the ϕ_n). The emission modes for $q = 1$, 2, and 3 are presented in the first, second and third row of Figure 2, respectively. It can be found that the phase repeats one, two, and three times from 0 to 2π upon one full circle around the center of the emitted light beam for $q = 1, 2$, and 3, respectively, which means the OAM can be designed on demand.

Now, we show that this OAM laser can naturally ensure the single-mode operation of desired mode. Here, we make a comparison between the modes with the same field distribution in the vertical cross section of the bend waveguide of the microring but with different field distributions in the azimuthal direction. For the standing-wave WGM, one mode (mode 1) of the pair of modes with $l = m$ is mainly confined in region 1, and the other mode (mode 2) is mainly confined in region 2. In addition, the cavity modes with $l \neq m$ all have nearly the same distribution weight in two regions. Thus, the desired mode 1 experiences the lower loss from region 2 than other modes, that is, enhancing the loss in region 2 can relatively suppress the lasing of other competing modes. In order to demonstrate this principle, we conduct the 3D full-wave simulations. We take the laser cavity with $q = 2$ and $m = 11$ as an example. Figure 3a shows the ratios of the volume integral of $|E|^2$ confined in region 1 to the volume integral of $|E|^2$ confined in region 2. Obviously, the electric fields of the desired mode 1 are mainly confined in region 1 ($\Gamma_1/\Gamma_2 \sim 4.2$). In contrast, the electric fields of mode 2 are mainly confined in region 2 ($\Gamma_1/\Gamma_2 \sim 1/4.2$). For other WGMs (for instance the mode pair with $l = 10$ and mode pair with $l = 12$.), electric fields are nearly equally distributed in region 1 and region 2 ($\Gamma_1/\Gamma_2 \sim 1$). Thus, as the loss in region 2 increases, the quality factor of mode 1 (red line in Figure 3b) decreases slower than other modes (blue line and black lines in Figure 3b). As a result, the single-mode lasing of the desired mode will be benefited when the perturbation loss of region 2 is relatively large. It is worth noting that the OAM natures of the emission mode will not change as the loss in region 2 increases as shown in Figure 3c.

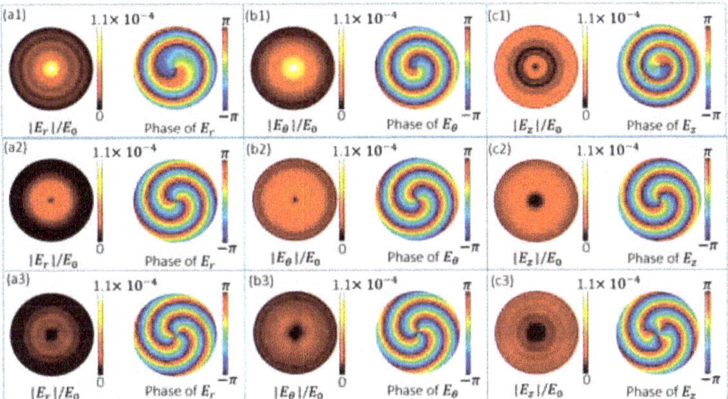

Figure 2. Tuning the OAM of the emitted light. The first row [(**a1**)–(**c1**)], second row [(**a2**)–(**c2**)], and third row [(**a3**)–(**c3**)] of the figure present the simulation results of the modes for q = 1, 2 and 3, respectively. These results are in the cross section (the radius of the area is 5000 nm) of the emission modes at 2 μm above the upper surface of the cavity. It is worth noting that the corresponding modes inside the cavity (mode 1) have nearly the same field intensity.

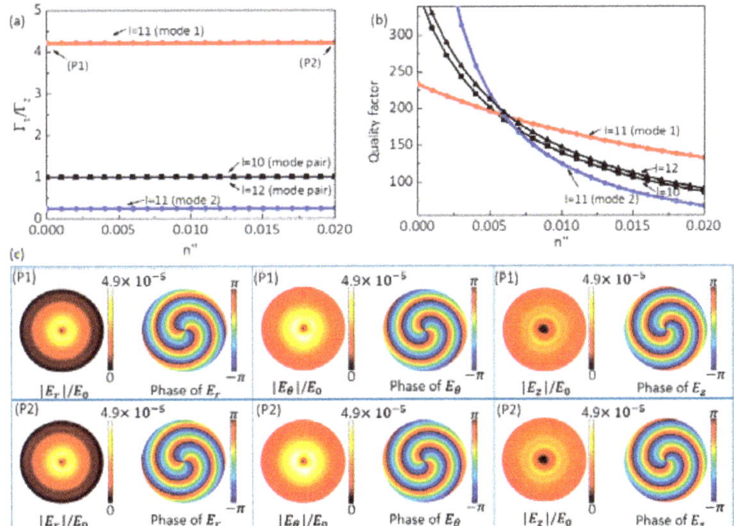

Figure 3. Suppressing the lasing of the competing modes. (**a**) The n'' dependence of the ratios of the volume integral of $|E|^2$ confined in the region 1 to the volume integral of $|E|^2$ confined in region 2, which is calculated as $\Gamma_1/\Gamma_2 = \iiint_{\text{Region 1}} |E|^2 dv / \iiint_{\text{Region 2}} |E|^2 dv$. Here, n'' is the modulation in Equation (2). (**b**) The quality factors of the modes as a function of the n''. (**c**) The intensity ($|E_r|/E_0$, $|E_\theta|/E_0$, and $|E_z|/E_0$) and phase distributions of the emission mode at P1 and P2 of the (**a**). The (**c**) are in the cross section (the radius of the area is 5000 nm) of the emission mode at 2 μm above the upper surface of the cavity. In this example, $q = 2$ and $m = 11$. The red lines correspond to the desired mode 1 with $l = 11$, the blue lines correspond to mode 2 with $l = 11$, the black lines correspond to the modes with $l = 10$, and 12.

Finally, we investigate the switch of the OAM of the emitted light. In order to control the OAM of the emitted light, we modulate the effective refractive index of region 1 to couple one standing-wave WGM (mode 1) to an OAM mode, and modulate the effective refractive index of region 2 to couple the other standing-wave WGM (mode 2) to the other OAM mode. Then, the losses of region 1 and region 2 are used to select the lasing mode between mode 1 and mode 2. Here, the effective refractive indexes in the ring are

$$n_G = \begin{cases} n_{\text{InGaAsP}} + \eta_1 i + \delta + \delta i + \delta e^{i\phi_n^1}, & \theta_1(n) - \frac{\Delta\theta}{2} \le \theta \le \theta_1(n) + \frac{\Delta\theta}{2} (n = 1,3,5\ldots 2m-1) \\ n_{\text{InGaAsP}} + \eta_1 i + \delta + \delta i + \delta e^{i(\phi_n^1+\pi)}, & \theta_1(n) - \frac{\Delta\theta}{2} \le \theta \le \theta_1(n) + \frac{\Delta\theta}{2} (n = 2,4,6\ldots 2m) \\ n_{\text{InGaAsP}} + \eta_2 i + \delta + \delta i + \delta e^{i\phi_n^2}, & \theta_2(n) - \frac{\Delta\theta}{2} \le \theta \le \theta_2(n) + \frac{\Delta\theta}{2} (n = 1,3,5\ldots 2m-1) \\ n_{\text{InGaAsP}} + \eta_2 i + \delta + \delta i + \delta e^{i(\phi_n^2+\pi)}, & \theta_2(n) - \frac{\Delta\theta}{2} \le \theta \le \theta_2(n) + \frac{\Delta\theta}{2} (n = 2,4,6\ldots 2m), \end{cases} \tag{3}$$

where $\Delta\theta = \pi/2m, \theta_1(n) = \pi n/m\ (n = 1, 2, 3, \ldots, 2m), \theta_2(n) = \pi n/m + \Delta\theta\ (n = 1, 2, 3, \ldots, 2m)$, $\phi_n^1 = q_1 n\pi/m\ (n = 1, 2, 3, \ldots, 2m)$, $\phi_n^2 = q_2 n\pi/m\ (n = 1, 2, 3, \ldots, 2m)$. The electric field $|E|$ of mode 1 and mode 2 are mainly localized in region 1 and region 2, respectively. Thus, the lasing of mode 1 or mode 2 can be selectively suppressed by tuning the additional material loss ($\eta_1 i$) in region 1 and the additional material loss ($\eta_2 i$) in region 2, which enables the switch of the emission mode between two different OAM modes. Now, we present the results of the 3D full-wave simulation. In our example, we used the parameters $m = 11$, $\delta = 0.006$, $q_1 = 2$, $q_2 = 3$. We assume that the parameters (η_1, η_2) are tuned from $(\eta_1 = 0, \eta_2 = 0.006)$ to $(\eta_1 = 0.006, \eta_2 = 0)$. The gain of the InGaAsP can be achieved when being pumped. The homogeneous pumping gain of the InGaAsP ring is mimicked by adding an additional background imaginary part Δi ($\Delta < 0$) to the effective refractive index of the InGaAsP ring. It is worth noting that we do not consider the effect of the specific gain spectrum of the gain material for the sake of simplicity. Figure 4a shows that the loss of mode 1 is compensated for before the loss of mode 2 with the increase of the homogeneous pumping gain of the InGaAsP for $(\eta_1 = 0, \eta_2 = 0.006)$. Thus, mode 1 will be the lasing mode. In this case, the phases of E_r, E_z and E_θ of the emitted lasing beam all repeat two times from 0 to 2π upon one full circle around the center of the emitted light beam (Figure 4c). After tuning the system parameters (η_1, η_2) from $(\eta_1 = 0, \eta_2 = 0.006)$ to $(\eta_1 = 0.006, \eta_2 = 0)$, the loss of mode 2 is compensated before the loss of mode 1 with the increase of the homogeneous pumping gain of the InGaAsP as shown by Figure 4b. Thus, the lasing mode switches from mode 1 to mode 2. In this case, the emitted lasing beam is switched to the other OAM mode as shown in Figure 4d, in which the phases of the E_r, E_z and E_θ all repeat three times from 0 to 2π upon one full circle around the center of the emitted light.

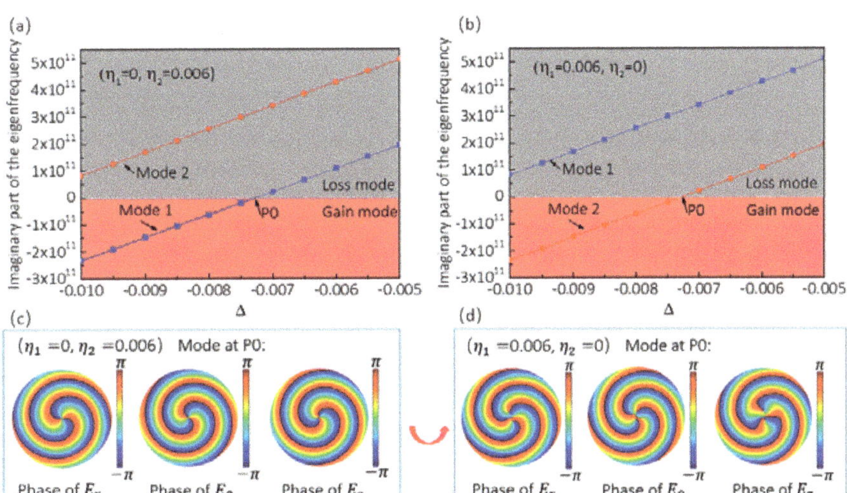

Figure 4. Switching the OAM carried by the emitted light. (**a**), (**b**), The homogeneous background gain dependence of the imaginary part of the eigenfrequencies of modes for the system with parameters ($\eta_1 = 0$, $\eta_2 = 0.006$) (**a**), and ($\eta_1 = 0.006$, $\eta_2 = 0$) (**b**). (**c**) The emitted lasing beam corresponding to P0 in (**a**). (**d**) The emitted lasing beam corresponding to P0 in (**b**). The phases of the emission modes in (**c**) and (**d**) are in the cross section (the radius of the area is 5000 nm) of the emission mode at 2 μm above the upper surface of the cavity.

3. Conclusions

In this paper, we propose a mechanism for constructing OAM light by modulating the scattered waves from the standing-wave WGM in the plasmonic ring cavity. Within this framework, the single-mode lasing can be realized. Moreover, the OAM of the emitted light can be designed on demand. This work provides a mechanism for exploring a single-mode OAM laser with tunable OAM at the microscale.

Author Contributions: Conceptualization, X.W.; writing—original draft, X.W.; writing—review and editing, X.W., X.H. and T.Z.; supervision, X.W., X.H. and T.Z. All authors have read and agreed to the published version of the manuscript.

Funding: This research was funded by National Natural Science Foundation of China (Grant Nos. 61822501, and 11734001), and the Beijing Natural Science Foundation (Grant No. Z180015).

Data Availability Statement: The data presented in this study are available within the article.

Conflicts of Interest: The authors declare no conflict of interest.

References

1. Allen, L.; Beijersbergen, M.W.; Spreeuw, R.J.C.; Woerdman, J.P. Orbital angular momentum of light and the transformation of Laguerre-Gaussian laser modes. *Phys. Rev. A* **1992**, *45*, 8185–8189. [CrossRef]
2. Bozinovic, N.; Yue, Y.; Ren, Y.; Tur, M.; Kristensen, P.; Huang, H.; Willner, A.E.; Ramachandran, S. Terabit-scale orbital angular momentum mode division multiplexing in fibers. *Science* **2013**, *340*, 1545–1548. [CrossRef] [PubMed]
3. Nicolas, A.; Veissier, L.; Giner, L.; Giacobino, E.; Maxein, D.; Laurat, J. A quantum memory for orbital angular momentum photonic qubits. *Nat. Photonics* **2014**, *8*, 234–238. [CrossRef]
4. Maiman, T.H. Stimulated Optical Radiation in Ruby Masers. *Nature* **1960**, *187*, 493–494. [CrossRef]
5. Khajavikhan, M.; Simic, A.; Katz, M.; Lee, J.H.; Slutsky, B.; Mizrahi, A.; Lomakin, V.; Fainman, Y. Thresholdless nanoscale coaxial lasers. *Nature* **2012**, *482*, 204–207. [CrossRef]
6. Lubatsch, A.; Frank, R. A Self-Consistent Quantum Field Theory for Random Lasing. *Appl. Sci.* **2019**, *9*, 2477. [CrossRef]
7. Lubatsch, A.; Frank, R. Quantum Many-Body Theory for Exciton-Polaritons in Semiconductor Mie Resonators in the Non-Equilibrium. *Appl. Sci.* **2020**, *10*, 1836. [CrossRef]

8. Okamoto, T.; Mori, M. Random Laser Action in Dye-Doped Polymer Media with Inhomogeneously Distributed Particles and Gain. *Appl. Sci.* **2019**, *9*, 3499. [CrossRef]
9. Zhang, W.X.; Xie, X.; Hao, H.M.; Dang, J.C.; Xiao, S.; Shi, S.S.; Ni, H.Q.; Niu, Z.C.; Wang, C.; Jin, K.J.; et al. Low-threshold topological nanolasers based on the second-order corner state. *Light Sci. Appl.* **2020**, *9*, 109. [CrossRef]
10. Schneider, C.; Rahimi-Iman, A.; Kim, N.Y.; Fischer, J.; Savenko, I.G.; Amthor, M.; Lermer, M.; Wolf, A.; Worschech, L.; Kulakovskii, V.D.; et al. An electrically pumped polariton laser. *Nature* **2013**, *497*, 348–352. [CrossRef]
11. Ye, Z.Y.; Su, M.; Li, J.N.; Jing, C.N.; Xu, S.B.; Liu, L.Q.; Ren, G.C.; Wang, X.L. Laser nano-technology of light materials: Precision and opportunity. *Opt. Laser Technol.* **2021**, *139*, 106988. [CrossRef]
12. Al-Shibaany, Z.Y.A.; Penchev, P.; Hedley, J.; Dimov, S. Laser Micromachining of Lithium Niobate-Based Resonant Sensors towards Medical Devices Applications. *Sensors* **2020**, *20*, 2206. [CrossRef]
13. Shen, Y.; Wang, X.J.; Xie, Z.W.; Min, C.J.; Fu, X.; Liu, Q.; Gong, M.; Yuan, X.C. Optical vortices 30 years on: OAM manipulation from topological charge to multiple singularities. *Light Sci. Appl.* **2019**, *8*, 90. [CrossRef]
14. Zeng, J.; Li, L.; Yang, X.; Gao, J. Generating and separating twisted light by gradient–rotation split-ring antenna metasurfaces. *Nano Lett.* **2016**, *16*, 3101–3108. [CrossRef]
15. Zeng, J.; Gao, J.; Luk, T.S.; Litchinitser, N.M.; Yang, X. Structuring light by concentric-ring patterned magnetic metamaterial cavities. *Nano Lett.* **2015**, *15*, 5363–5368. [CrossRef] [PubMed]
16. Kotlyar, V.V.; Almazov, A.A.; Khonina, S.N.; Soifer, V.A.; Elfstrom, H.; Turunen, J. Generation of phase singularity through diffracting a plane or Gaussian beam by a spiral phase plate. *J. Opt. Soc. Am. A* **2005**, *5*, 849–861. [CrossRef] [PubMed]
17. Khonina, S.N.; Podlipnov, V.V.; Karpeev, S.V.; Ustinov, A.V.; Volotovsky, S.G.; Ganchevskaya, S.V. Spectral control of the orbital angular momentum of a laser beam based on 3D properties of spiral phase plates fabricated for an infrared wavelength. *Opt. Express* **2020**, *12*, 18407–18417. [CrossRef]
18. Li, H.L.; Phillips, D.B.; Wang, X.Y.; Ho, Y.-L.D.; Chen, L.; Zhou, X.Q.; Zhu, J.B.; Yu, S.Y.; Cai, X.L. Orbital angular momentum vertical-cavity surface-emitting lasers. *Optica* **2015**, *2*, 547–552. [CrossRef]
19. Padgett, M.; Courtial, J.; Allen, L. Light's orbital angular momentum. *Phys. Today* **2004**, *57*, 35–40. [CrossRef]
20. Marrucci, L.; Manzo, C.; Paparo, D. Optical spin-to-orbital angular momentum conversion in inhomogeneous anisotropic media. *Phys. Rev. Lett.* **2006**, *96*, 163905. [CrossRef] [PubMed]
21. Yu, N.; Genevet, P.; Kats, M.; Aieta, F.; Tetienne, J.; Capasso, F.; Gaburro, Z. Light propagation with phase discontinuities: Generalized laws of reflection and refraction. *Science* **2011**, *334*, 333–337. [CrossRef]
22. Yang, Y.; Wang, W.; Moitra, P.; Kravchenko, I.I.; Briggs, D.P.; Valentine, J. Dielectric meta-reflectarray for broadband linear polarization conversion and optical vortex generation. *Nano Lett.* **2014**, *14*, 1394–1399. [CrossRef] [PubMed]
23. Heckenberg, N.R.; McDuff, R.; Smith, C.P.; White, A.G. Generation of optical phase singularities by computer-generated holograms. *Opt. Lett.* **1992**, *17*, 221–223. [CrossRef] [PubMed]
24. Senatsky, Y.; Bisson, J.; Li, J.; Shirakawa, A.; Thirugnanasambandam, M.; Ueda, K. Laguerre-Gaussian modes selection in diode-pumped solid-state lasers. *Opt. Rev.* **2012**, *19*, 201–221. [CrossRef]
25. Oron, R.; Davidson, N.; Friesem, A.A.; Hasman, E. Efficient formation of pure helical laser beams. *Opt. Commun.* **2000**, *182*, 205–208. [CrossRef]
26. Oron, R.; Danziger, Y.; Davidson, N.; Friesem, A.A.; Hasman, E. Laser mode discrimination with intra-cavity spiral phase elements. *Opt. Commun.* **1999**, *169*, 115–121. [CrossRef]
27. Bisson, J.-F.; Senatsky, Y.; Ueda, K. Generation of Laguerre-Gaussian modes in Nd: YAG laser using diffractive optical pumping. *Laser Phys. Lett.* **2005**, *2*, 327–333. [CrossRef]
28. Okida, M.; Omatsu, T.; Itoh, M.; Yatagai, T. Direct generation of high power Laguerre-Gaussian output from a diode-pumped Nd: YVO 4 1.3-μm bounce laser. *Opt. Express* **2007**, *15*, 7616–7622. [CrossRef]
29. Caley, A.J.; Thomson, M.J.; Liu, J.; Waddie, A.J.; Taghizadeh, M.R. Diffractive optical elements for high gain lasers with arbitrary output beam profiles. *Opt. Express* **2007**, *15*, 10699–10704. [CrossRef]
30. Ito, A.; Kozawa, Y.; Sato, S. Generation of hollow scalar and vector beams using a spot-defect mirror. *J. Opt. Soc. Am. A* **2010**, *27*, 2072–2077. [CrossRef] [PubMed]
31. Kano, K.; Kozawa, Y.; Sato, S. Generation of purely single transverse mode vortex beam from a He-Ne laser cavity with a spot-defect mirror. *Int. J. Opt.* **2011**, *2012*, 359141. [CrossRef]
32. Naidoo, D.; Roux, F.S.; Dudley, A.; Litvin, I.; Piccirillo, B.; Marrucci, L.; Forbes, A. Controlled generation of higher-order Poincaré sphere beams from a laser. *Nat. Photon.* **2016**, *10*, 327–332. [CrossRef]
33. Al-Attili, A.Z.; Burt, D.; Li, Z.; Higashitarumizu, N.; Gardes, F.Y.; Oda, K.; Ishikawa, Y.; Saito, S. Germanium vertically light-emitting micro-gears generating orbital angular momentum. *Opt. Express* **2018**, *26*, 34675–34688. [CrossRef] [PubMed]
34. Miao, P.; Zhang, Z.; Sun, J.; Walasik, W.; Longhi, S.; Litchinitser, N.M.; Feng, L. Orbital angular momentum microlaser. *Science* **2016**, *353*, 464–467. [CrossRef] [PubMed]
35. Feng, L.; Ayache, M.; Huang, J.; Xu, Y.-L.; Lu, M.-H.; Chen, Y.-F.; Fainman, Y.; Scherer, A. Nonreciprocal light propagation in a silicon photonic circuit. *Science* **2011**, *333*, 729–733. [CrossRef]

Article

High-Performance Flexible Transparent Electrodes Fabricated via Laser Nano-Welding of Silver Nanowires

Tao Wang [1,†], Yinzhou Yan [2,*,†], Liye Zhu [2], Qian Li [2], Jing He [2], Xiaoxia Zhang [1], Xi Li [2], Xiaohua Zhang [2], Yongman Pan [2] and Yue Wang [1]

[1] Faculty of Science, Beijing University of Technology, Beijing 100124, China; wangtao@emails.bjut.edu.cn (T.W.); xxzhang2021@163.com (X.Z.); wy2001@bjut.edu.cn (Y.W.)
[2] Faculty of Materials and Manufacturing, Beijing University of Technology, Beijing 100124, China; zly2019@emails.bjut.edu.cn (L.Z.); liqianwwh@163.com (Q.L.); S201913010@emails.bjut.edu.cn (J.H.); Lisimeng@emails.bjut.edu.cn (X.L.); zxh2636104711@163.com (X.Z.); panym@emails.bjut.edu.cn (Y.P.)
* Correspondence: yyan@bjut.edu.cn
† Equal contribution to this work.

Abstract: Silver nanowires (Ag-NWs), which possess a high aspect ratio with superior electrical conductivity and transmittance, show great promise as flexible transparent electrodes (FTEs) for future electronics. Unfortunately, the fabrication of Ag-NW conductive networks with low conductivity and high transmittance is a major challenge due to the ohmic contact resistance between Ag-NWs. Here we report a facile method of fabricating high-performance Ag-NW electrodes on flexible substrates. A 532 nm nanosecond pulsed laser is employed to nano-weld the Ag-NW junctions through the energy confinement caused by localized surface plasmon resonance, reducing the sheet resistance and connecting the junctions with the substrate. Additionally, the thermal effect of the pulsed laser on organic substrates can be ignored due to the low energy input and high transparency of the substrate. The fabricated FTEs demonstrate a high transmittance (up to 85.9%) in the visible band, a low sheet resistance of 11.3 Ω/sq, high flexibility and strong durability. The applications of FTEs to 2D materials and LEDs are also explored. The present work points toward a promising new method for fabricating high-performance FTEs for future wearable electronic and optoelectronic devices.

Keywords: flexible transparent electrodes; silver nanowires; laser nano-welding; organic electronics

Citation: Wang, T.; Yan, Y.; Zhu, L.; Li, Q.; He, J.; Zhang, X.; Li, X.; Zhang, X.; Pan, Y.; Wang, Y. High-Performance Flexible Transparent Electrodes Fabricated via Laser Nano-Welding of Silver Nanowires. *Crystals* **2021**, *11*, 996. https://doi.org/10.3390/cryst11080996

Academic Editor: Giancarlo Salviati

Received: 7 August 2021
Accepted: 18 August 2021
Published: 21 August 2021

1. Introduction

High-performance flexible transparent electrodes (FTEs) with outstanding mechanical and optical properties facilitate the rapid development of wearable electronics and optoelectronics [1]. Compared with rigid materials such as silicon and silica, elastic substrates make the devices foldable, twistable, compressible, and stretchable without compromising stability and reliability, allowing for a wide range of applications, including flexible electronic displays, organic light-emitting diodes (OLEDs), solar cells, and electronic skins [2–5]. FTEs are critical components in the above-mentioned wearable devices to ensure a high-efficiency power supply with low energy consumption. Moreover, the excellent optical transmittance would replace indium tin oxide (ITO) for next-generation, highly flexible optoelectronic applications, overcoming the drawbacks of time-consuming synthesis, indium requirement, and fragility during stretching and bending. In recent decades, numerous efforts have been made to develop alternative FTEs, including graphene [6–8], nanowires [9–12], carbon nanotubes (CNTs) [13–15], and PEDOT:PSS [16–18].

Silver nanowires (Ag-NWs) have attracted considerable attention as FTE materials owing to their good electrical conductivity, high transparency in the visible band, excellent ductility, and facile fabrication process [19]. In particular, the conductive networks formed by ultralong metallic nanofibers have shown promise for application to FTEs because of their superior optical, electrical, and mechanical properties [20,21]. A variety of electronic components with Ag-NW FTEs were fabricated in previous studies, including flexible and transparent antennas [22], wireless circuits [23], transistors [24], and sensors [25], demonstrating the potential applications in wearable devices. However, the contact resistance between Ag-NWs and adhesion to the organic substrates have presented two major challenges for high-performance FTEs. Methods for the improvement of contact resistances have been sophisticatedly presented in previous works. Thermal annealing has received the most attention due to its facile fabrication process and high welding quality. However, thermal annealing requires an oxygen-free environment due to the high chemical reactivity of Ag-NWs at high temperature. Moreover, thermal annealing is not suitable for the flexible organic substrates due to the low melting temperature of organic materials. Consequently, a low-temperature thermal annealing (below 200 °C) was developed [26]. Other physical welding methods employing force, electricity, light, etc. were also developed to realize Ag-NW network welding without significant thermal effects for FTEs [27–35]. Although high-pressure welding optimized the conductivity of the Ag-NW FTEs down to 8.4 Ω/sq, the substrates and other layers could be destroyed [27]. To solve this problem, Liu et al. proposed a capillary-force-induced cold-welding technique in which the moisture-treated Ag-NWs exhibited a significant reduction in sheet resistance (~37 Ω/sq) due to the giant capillary forces exerted between Ag-NWs [28]. Unfortunately, the poor adhesion of Ag-NWs on the substrate has not been improved. Another method that has been used is electric welding; it is based on the high contact resistance at the junction of Ag-NWs, where Joule heating caused by the application of bias voltages results in localized melting, welding the Ag-NWs [29]. The main drawback of this method is the non-homogeneous current distribution due to the complex Ag-NW network, which proved unsuitable for large-area welding. Hong et al. achieved Ag-NW nano-welding by electron-beam irradiation, which provided sufficient energy to melt the Ag-NWs [30]. However, the required vacuum chamber and the high-cost instrument limited the method for industrial applications. Compared with electron beams, light is an ideal energy source to induce the nano-welding of Ag-NWs. Liang et al. employed ultraviolet A (UVA) light irradiation for Ag-NW nano-welding, reducing the sheet resistance by three orders of magnitude (down to 25 Ω/sq) with a good transparency (97%) [31]. The femtosecond pulsed laser (fs-pulsed laser) was also used to irradiate Ag-NWs, where the excited localized surface plasmon resonances (LSPRs) at the gaps between Ag-NWs generated a considerable enhancement of electric field strength, inducing local melting for nano-welding. Meanwhile, the polyethylene glycol terephthalate (PET) substrate was not damaged during the laser irradiation. The obtained sheet resistance and optical transmittance by fs-pulsed nano-welding were 16.1 Ω/sq and 91%, respectively [32]. Although nanosecond pulsed lasers (ns-pulsed lasers) were also tested for the nano-welding of Ag-NWs [33], most of the laser-irradiation methods were only suitable for inorganic and limited organic substrates with high melting temperatures and stiffnesses (glass, silica, PET, PVA), owing to the requirement of thermal endurance from laser energy input. Generally, the most used highly flexible and bio-compatible organic materials (e.g., polydimethylsiloxane, PDMS), possess low melting points and stiffnesses. Therefore, a nano-welding technique for junction-localized energy confinement that will not deteriorate the substrate is still lacking. In addition, the adhesion of Ag-NW networks to the organic substrate also needs to be resolved by jointing Ag-NWs with the substrate gently [9,10].

In this work, we developed a facile technique for fabricating high-performance FTEs via ns-pulsed laser nano-welding of Ag-NWs on organic substrates, creating Ag-NWs/PMMA/PDMS sandwich structures. LSPR-induced laser energy confinement at the junctions between the Ag-NWs was employed to generate a high temperature for nano-welding in order to reduce contact resistances while dramatically boosting the electrical conductivity of the Ag-NWs. Meanwhile, the Ag-NW networks were jointed with the organic substrate for good adhesive strength. The ns-pulsed laser energy for nano-welding was low enough to avoid thermal damage to the organic substrates completely. The fabricated FTEs in this work demonstrated superior mechanical, electrical, and optical properties. The compatibility of FTEs with 2D materials and traditional LEDs was also explored for the design of flexible optoelectronic devices in the future.

2. Materials and Methods

A schematic of the procedure for fabricating Ag-NW FTEs is illustrated in Figure 1a. The glass substrates (2 cm × 2 cm) were first cleaned ultrasonically in acetone, isopropanol, and deionized water, successively, for 5 min. The polymethyl methacrylate (PMMA) solution in toluene (purchased from Shanghai Kexinda Polymer Materials Co., Ltd., Shanghai, China) was spin-coated onto the glass substrate at 8000 rpm for 40 s to achieve a thickness of ~10 μm, then cured in ambient atmospheric for 10 s. Then, the PMMA film was treated with plasma (YZD08-5C, 80 W, Saiaote Technology Co. Ltd., Wuhu, China) for 30 s to achieve a hydrophilic surface, beneficial for both uniform spreading and the prevention of agglomeration during Ag-NW suspension deposition. The Ag-NWs (purchased from Nanjing XFNANO Co., Ltd., Nanjing, China), with a mean length of ~100 μm and diameter of 30.0 ± 5.0 nm, diluted in isopropyl alcohol (IPA) for various concentrations (0.3–0.7 mg/mL), were sprayed onto the PMMA surfaces. To minimize the coffee-ring effect on Ag-NWs deposition, 3M low-adhesion tapes were employed to confine the suspension flow within a specific region. The homogeneous Ag-NWs were therefore formed after the suspension was dried. The densities of Ag-NWs on PMMA films were controlled by two synthesis parameters: the concentration of the Ag-NWs suspension (0.3–0.7 mg/mL) and the number of spraying layers (5–20 layers). Afterwards, the Ag-NWs/PMMA film was irradiated by a 532 nm ns-pulsed laser (Spectra-Physics, LAB-190-30H, 10 ns, 27 Hz) under ambient conditions at room temperature, achieving Ag-NW nano-welding and jointing with the PMMA substrate. It should be noted that the highly transparent PMMA with respect to 532 nm (95%) and the short pulse duration minimized the thermal damage on the substrate within the irradiation time of 50 s. The oxidation of Ag-NWs during laser irradiation was also negligible. The PDMS Sylgard 184 (purchased from Dow corning Co., Ltd., Midland, MI, USA) was mixed and stirred with the curing agent at the ratio of 10:1 by weight to obtain the PDMS colloidal solution. The fabricated Ag-NWs/PMMA film was then flipped over on the spinner. The PDMS solution was spin-coated onto the bottom side of the Ag-NWs/PMMA film at a rotation speed of 5000 rpm for 30 s. The PDMS film was baked at 120 °C for 2 min on a plate heater in a vacuum chamber for solidification and bubble removal. The Ag-NWs/PMMA/PDMS film was therefore obtained, where the thickness of the PDMS film was ~20 μm. It should be noted that the PDMS film provided an outstandingly flexible substrate for the FTEs, and the plasma-treated PMMA film served as a buffer layer for high adhesion of Ag-NWs and PDMS in the sandwich structure, as shown in Figure 1b. Figure 1c demonstrates the optical and mechanical performance of the Ag-NWs/PMMA/PDMS FTEs. In addition, the MoS_2 monolayer (purchased from Shenzhen Six-Carbon Technology, Shenzhen, China), grown on an *n*-type Si substrate with a 300 nm-thick SiO_2 film, was also employed for future experiments.

The sheet resistances were measured using the four-point probe method (Suzhou Jingge Electronic Co., Ltd., ST2258C, Suzhou, China), in which the mean values of sheet resistances from six random points of the FTEs were calculated. The optical transmittances of the FTEs were obtained by a UV-VIS spectrophotometer (Shimadzu, UV-3600, Kyoto, Japan). The morphologies of the Ag-NWs in the FTEs were captured by optical microscopy (Olympus BX-51, Tokyo, Japan), scanning electron microscopy (Hitachi, SU8220, Tokyo, Japan), and transmission electron microscopy (FEI, Tecnai G2-20-S-TWIN, Lausanne, Switzerland). The film thickness was acquired by a profilometer (Veeco Dektak-XT, Bruker, Billerica, MA, USA). The gate-voltage applied to the MoS_2 was supplied by a DC power supply (Beijing leading Hongzhi Electronic Technology Co., Ltd., XD1715A-120, Beijing, China). The photoluminescence spectra were analyzed by a SmartRaman confocal-micron-Raman system (developed by Institute of Semiconductors, CAS, Beijing, China) with a 10x/*NA*0.25 objective lens (Olympus, MPlan N, Tokyo, Japan) under the backscattering geometry, which was coupled with a Horiba LabRam iHR550 spectrometer (Kyoto, Japan) with a 100 lines/mm grating and a CCD detector. The excitation CW laser wavelength was 633 nm (HNL 100-EC-PS, 25.8 μW, Changchun New Industries Technology Co., Ltd., Changchun, China). The electrical power supply and current measurement for LEDs were provided by a source meter (Keithley 4200-SCS, Tektronix, Beaverton, OR, USA).

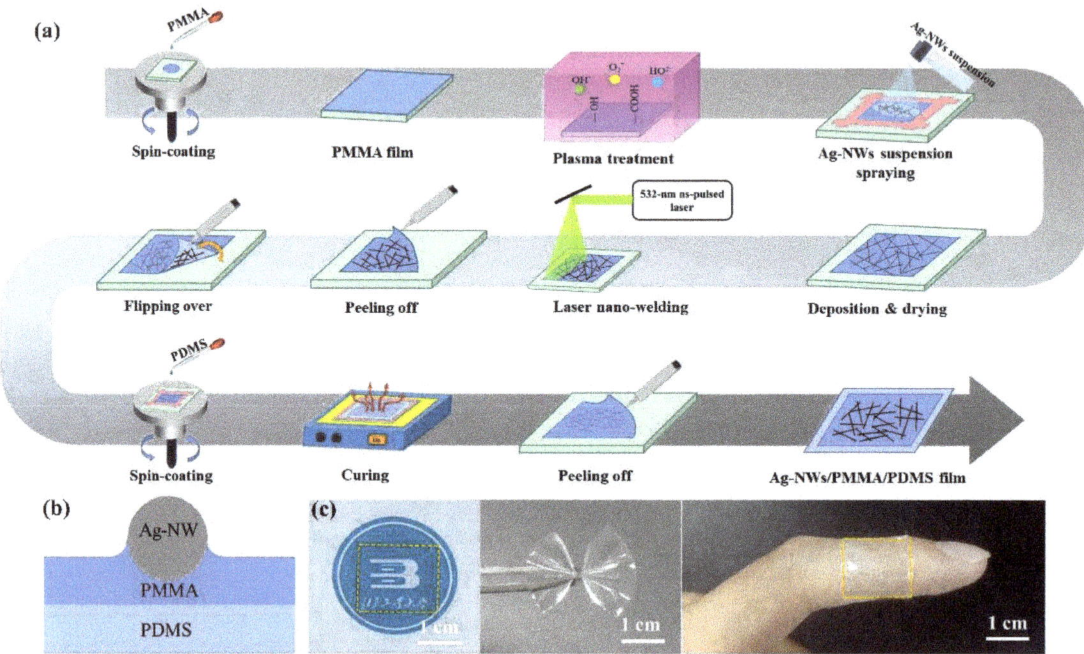

Figure 1. Fabrication of Ag-NWs/PMMA/PDMS FTEs. (**a**) Schematic of synthesis procedure. (**b**) Cross-section schematic of the FTEs. (**c**) Mechanical and optical properties of the FTEs.

The numerical simulation was performed using the finite element algorithm method in the COMSOL Multiphysics (licensed by COMSOL Co., Ltd., Stockholm, Sweden) software package. A 2D cross-sectional model was developed to calculate the electric fields regulated at the cross-junctions and gaps between Ag-NWs. The Ag-NW diameter was 30 nm, and the relative permittivity was $-9.3751 + 0.83203i$ according to Drude's model. The ambient environment was set as air. For the analysis of electric field enhancement, a plane wave with a wavelength of 532 nm was incident onto the Ag-NWs. Perfectly matched layers were applied as the boundary conditions.

3. Results and Discussion

3.1. Morphology of Laser Nano-Welded Ag-NWs

The morphologies of Ag-NWs nano-welded via various laser fluences are shown in Figure 2. The as-deposited Ag-NWs were randomly distributed on the PMMA film, forming a conductive network, as shown in Figure 2a. The close-up view in Figure 2b further demonstrates the cross junctions of the Ag-NWs before laser nano-welding. The contacts between Ag-NWs and with the PMMA substrate were ascribed to van der Waals forces. The fluence threshold for laser nano-welding was ~10.0 mJ/cm^2, by which the cross junctions were melted slightly and connected after resolidification. The melting phenomenon became obvious with the laser fluence increasing to 17.4 mJ/cm^2. It can be clearly seen in Figure 2c–e that only the cross junctions were melted during laser irradiation whereas the other parts of Ag-NWs were not affected, indicating the laser-induced thermal effect was confined to the contact points. When the laser fluence was greater than 27.9 mJ/cm^2, the Ag-NWs sustained thermal damage. The high molten volumes and surface tensile effect broke the Ag-NWs, as shown in Figure 2f. Figure 2g,h further exhibits the welded points before and after laser irradiation, providing strong evidence of Ag-NW melting at the cross junction. To further optimize the laser nano-welding parameters, the effects of laser fluence and irradiation time on sheet resistance were studied, as shown in Figure 2i. The sheet resistance of as-deposited Ag-NWs was 110 Ω/sq and dramatically reduced via laser nano-welding within tens of seconds. The increased laser fluence and irradiation time both lowered the sheet resistance. It can also be seen that the sheet resistances were close to constants dominated by laser fluences as the irradiation time exceeded 50 s. The irradiation time of 50 s was therefore chosen as the optimal parameter to avoid oxidation of the Ag-NWs and thermal damages on the organic substrate. It should be noted that the high sheet resistance with a laser fluence of 37.9 mJ/cm^2 at an irradiation time of 50 s was due to the deterioration of the Ag-NWs, as shown in Figure 2f. Hence, the laser fluence was set to 27.9 mJ/cm^2 with an irradiation of 50 s for the lowest sheet resistance down to 11.3 Ω/sq, whereby the thermal effect on organic substrates was also negligible.

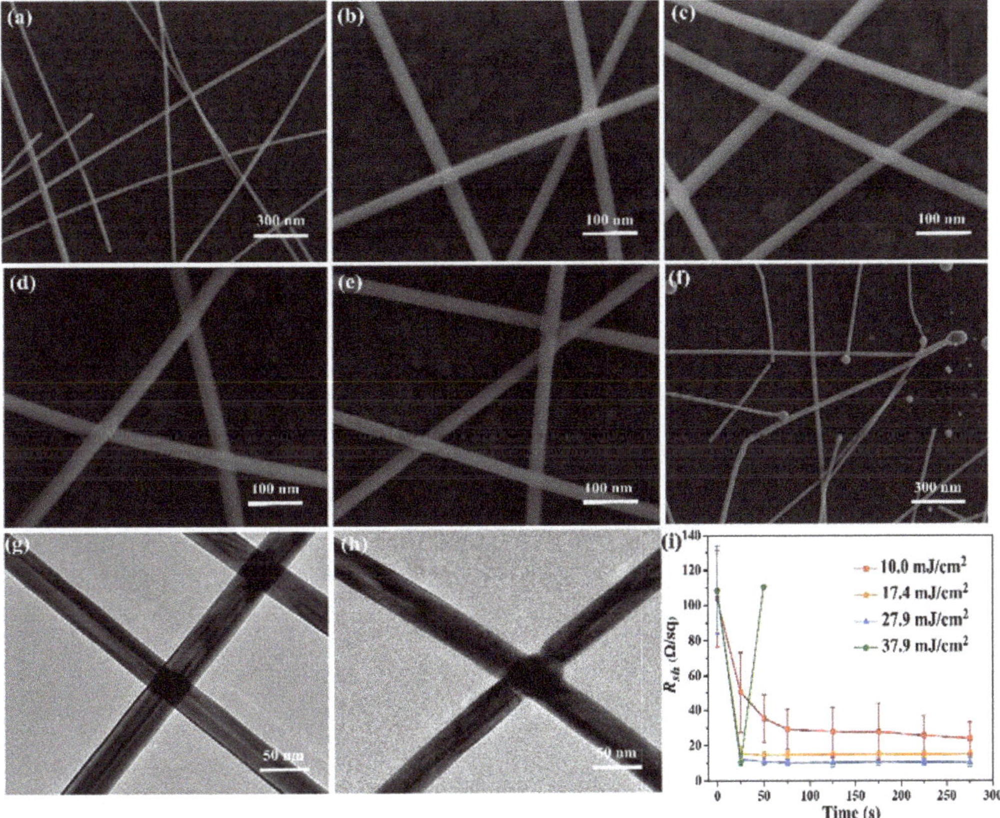

Figure 2. Morphologies and electrical properties of Ag-NWs/PMMA/PMMA FTEs via ns-pulsed laser nano-welding under various process parameters. (**a**) SEM images. (**b**) Close-up view of as-deposited Ag-NWs before nano-welding. (**c–f**) Laser nano-welding of Ag-NWs under laser fluence of (**c**) 10.0 mJ/cm^2, (**d**) 17.4 mJ/cm^2, (**e**) 27.9 mJ/cm^2, and (**f**) 37.9 mJ/cm^2 at irradiation time of 50 s. (**g,h**) TEM images of (**g**) as-deposited and (**h**) laser nano-welded Ag-NWs. (**i**) Evolution of sheet resistance with laser fluence and irradiation time.

3.2. Mechanism of Laser Nano-Welding of Ag-NWs on PMMA

To further study the mechanism of laser nano-welding of the Ag-NWs on organic substrates, a numerical simulation of laser interaction with Ag-NWs was performed. Two typical structures of as-deposited Ag-NWs were considered, as illustrated in Figure 3a,b, including cross-junction and head-to-head configurations. It is well-acknowledged that the LSPRs in the vicinity of Ag-NWs can be excited by light with a wavelength of ~50 nm, by which the electromagnetic (EM) fields are significantly enhanced and localized around the Ag-NWs. Meanwhile, the Joule heating caused by ohmic energy loss resulted in melting for the nano-welding of Ag-NWs [36–39]. The electric fields in the two structures are shown in Figure 3c,d. It can be clearly observed that the electrical intensities were boosted in the gaps between Ag-NWs, where the enhancement ratios were 573.4 and 2193.2 for cross-junction and head-to-head structures, respectively. As a result, these regions were first melted under laser irradiation and connected, realizing Ag-NW welding. The electrical intensities far from these gaps were extremely low, preventing the deterioration of Ag-NWs. The numerical simulation was in agreement with the experimental results, revealing the mechanism of laser nano-welding of the Ag-NWs. Furthermore, selective nano-welding was conducive to Ag-NW network jointing with the PMMA. Owing to its low melting point (~200 °C), the PMMA film can be melted due to the high temperature at the Ag-NW welding regions, and the Ag-NW network was partially embedded into the organic film for good adhesion. To validate the hypothesis, a 3M low-adhesion tape was employed to press and remove the Ag-NWs from the FTEs, as shown in Figure 3e,f. The adhesion between Ag-NWs and PMMA was significantly strengthened after laser nano-welding, as shown by the slight reduction of Ag-NW quantity under removal cycles of over 100. This confirmed the jointing of the Ag-NW network and PMMA substrate by laser irradiation.

3.3. Synthesis Optimization of Ag-NWs/PMMA/PDMS FTEs

The criteria for high-quality FTEs consist of electrical conductivity and transmittance. The ideal FTEs should possess high conductivity, good transmittance, and durable flexibility. The transmittance was generally reduced as conductivity increased due to light scattering and absorption by the high concentration of Ag-NWs. Therefore, the balance between conductivity and transmittance should be maintained by optimizing the density of Ag-NWs on the PMMA film. The multiple-spraying strategy was therefore employed to control the density by two parameters, that is, the concentration of the Ag-NWs suspension and the number of spraying layers. To obtain a stable sheet resistance, the number of spraying layers was greater than five. Figure 4 shows the sheet resistances and transmittances of Ag-NWs/PMMA/PDMS FTEs under different suspension concentrations and spraying layers after laser nano-welding with 28.9 mJ/cm^2 for 50 s. In Figure 4a–e, it can be clearly seen that the sheet resistance and transmittance were both decreased as the concentration and number of spraying layers increased. For the low concentration of 0.3 mg/mL, the sheet resistance was dramatically reduced when the number of spraying layers was eight, and then kept constant. Furthermore, the transmittance was linearly reduced from 89.9% to 63.4% as the number of spraying layers and concentration increased. The higher concentration could reduce the number of spraying layers down to five for the lowest sheet resistances. To evaluate the optimal performance in optical and electrical properties, the figure of merit (*FoM*) was employed as follows [40]:

$$FoM = \frac{T^{10}}{R_{sh}} \quad (1)$$

where T is the transmittance at 550 nm and R_{sh} is the sheet resistance of the FTEs. The optimal *FoMs* under various suspension concentrations were extracted and are plotted in Figure 4f. The suspension concentration of 0.5 mg/mL with five spraying layers was selected for the highest *FoM*, where the transmittance was 85.9% and the sheet resistance was 11.3 Ω/sq. The performance was comparable with that obtained in previous works [27–35].

The advantages of the technique developed in this work are its facile fabrication method and the low cost of FTE synthesis.

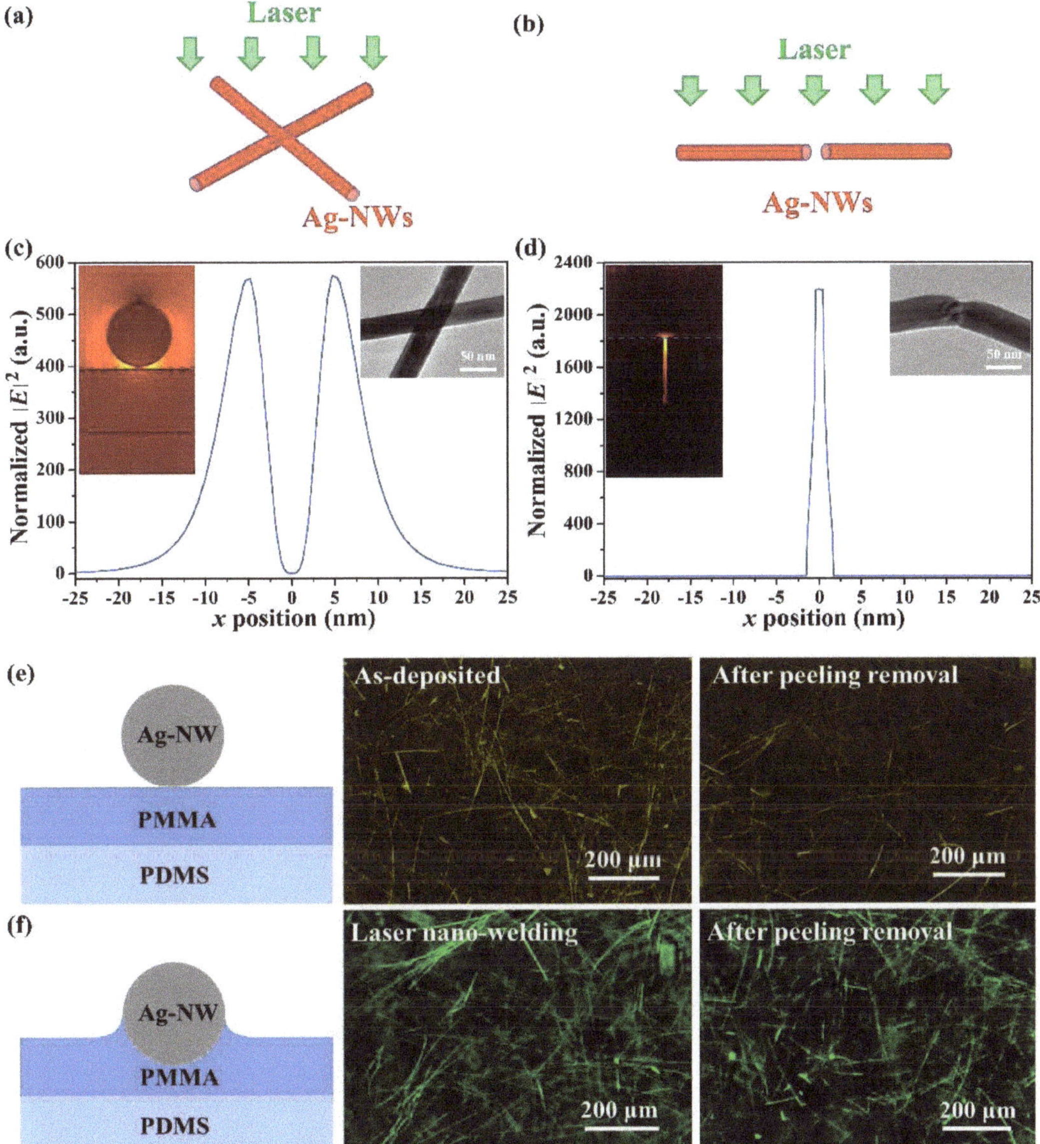

Figure 3. Numerical simulation of laser nano-welding of Ag-NWs and validation of Ag-NWs jointing with PMMA substrates. (**a**) Cross junction. (**b**) Head-to-head configurations for simulation. (**c**,**d**) Electric field distributions and intensities (**c**) near the cross junction between Ag-NWs and (**d**) near the gap of head-to-head Ag-NWs. (**e**,**f**) Removal experiments using 3M low-adhesion tapes for (**e**) as-deposited and (**f**) laser nano-welded Ag-NWs on PMMA film.

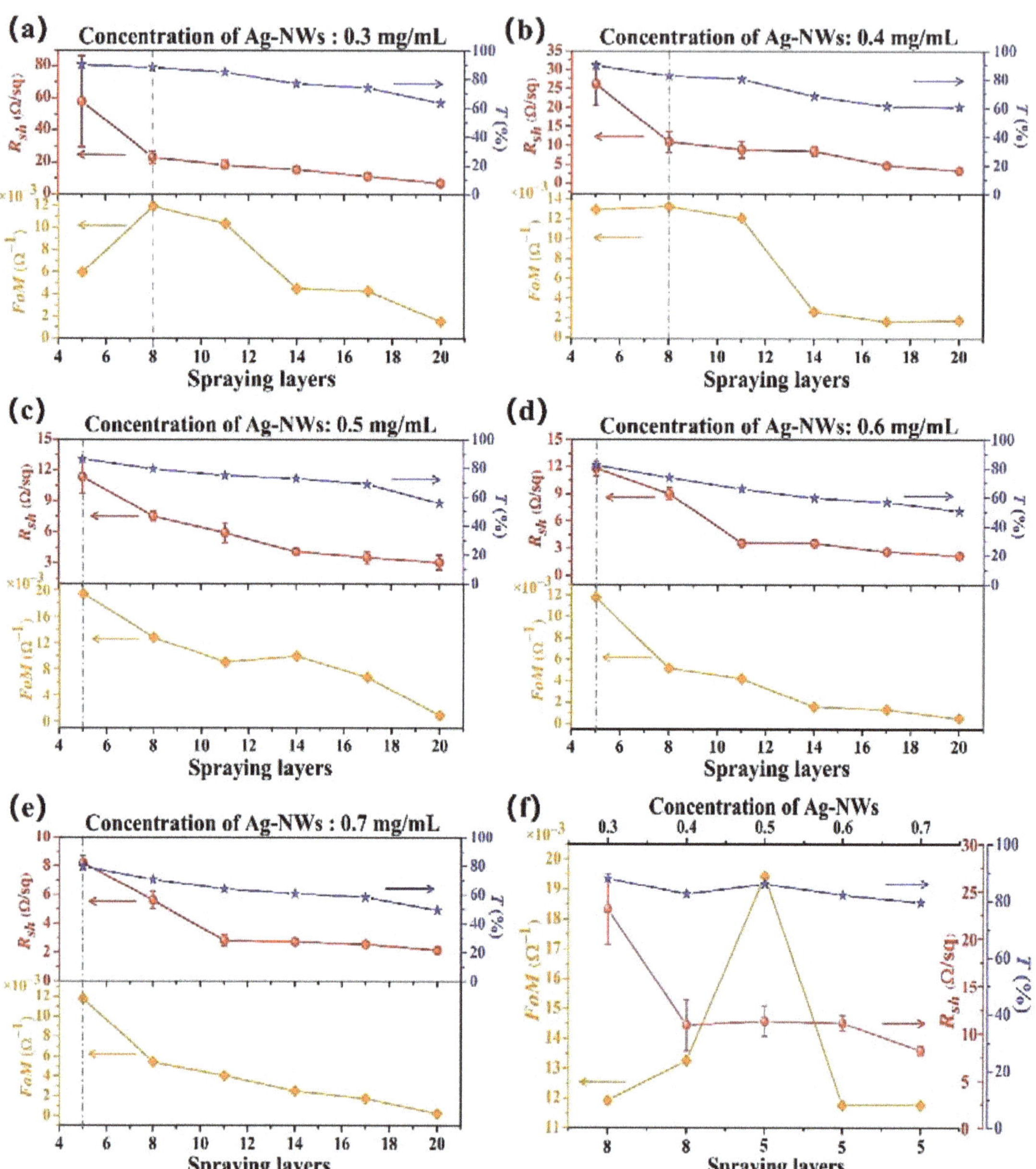

Figure 4. Evolution of transmittance and sheet resistance of Ag-NWs/PMMA/PDMS FTEs with the number of spraying layers using Ag-NW suspension concentrations of (**a**) 0.3 mg/mL, (**b**) 0.4 mg/mL, (**c**) 0.5 mg/mL, (**d**) 0.6 mg/mL, and (**e**) 0.7 mg/mL, as well as the calculated *FoM*s. (**f**) Optimization from various suspension concentration and spraying layer parameter sets.

3.4. Durability of Ag-NWs/PMMA/PDMS FTEs

Figure 5a shows the transmittance spectra of PDMS, PMMA, and Ag-NWs/PMMA/PDMS FTEs in the visible band. Although the Ag-NWs slightly reduced the transmittance, the flat spectrum in the visible band indicated good transparency without wavelength selection, as shown in Figure 1c. The flexibility of Ag-NWs/PMMA/PDMS FTEs was examined by the fold-bending test. The fold-bending-induced strain (ε) was 75.6%, which was estimated by [41]

$$\varepsilon = \frac{t}{2R} \times 100\% \tag{2}$$

where t is the thickness of the FTE and R is the bending radius, which were 31.0 μm and 20.5 μm measured by the profilometer, respectively. Figure 5b shows the sheet resistance maintained for up to 100 cycles of fold-bending with high strain, and then increased due to the fatigue of Ag-NWs. In addition, the adhesion of Ag-NWs on the organic substrate was also investigated. The above-mentioned discussion indicates that the Ag-NW network was jointed onto the PMMA film during laser nano-welding, whereby the adhesion between Ag-NWs and substrate was significantly strengthened since the Ag-NWs were partially embedded in the PMMA film. The 3M low-adhesion tape was thereby used to carry out the Ag-NW adhesion test by pressing it onto the film and peeling it off. Figure 5c indicates that the sheet resistance could be kept constant for the peeling-off process for up to 110 cycles, confirming the good adhesion of Ag-NWs on the PMMA/PDMS film via laser nano-welding. Figure 5d exhibited the stability of the Ag-NWs/PMMA/PDMS FTEs in ambient atmospheric at room temperature, where the sheet resistances were constant for 1 month. The excellent durability, flexibility, adhesion, and period stability recommend the laser nano-welded Ag-NWs/PMMA/PDMS FTEs for wide application in future wearable devices.

3.5. Applications of Ag-NWs/PMMA/PDMS Films as High-Performance FTEs

To demonstrate the performance of Ag-NWs/PMMA/PDMS FTEs, two typical configurations of wearable devices were employed and fabricated as shown in Figure 6. The applications of 2D materials in flexible electronic/optoelectronic components have been well-acknowledged [42]. For vertical device structures, FTEs became important for 2D material wearable designs. Figure 6a shows the typical sandwich structure for gate-controlled MoS_2 luminescence, in which an Ag-NWs/PMMA/PDMS FTE covered an MoS_2 monolayer, which was deposited on a 300 nm thick SiO_2 isolayer oxidized from an n^+-Si substrate. The PMDS side was contacted with the MoS_2 monolayer for applying the gate voltage, U, from the Ag-NW network to the n^+-Si substrate. The generated electric field, E, was estimated by $E = U/d$, where d is the combined thickness of the PMMA, the PDMS, the MoS_2 monolayer, and the SiO_2 isolayer. The luminescence of the MoS_2 monolayer is regulated by the gate voltage due to the interaction of excitons with charge carriers via the phase-space filling effect [43,44]. This optoelectronic property can be used to realize electro-optical modulators operating in the visible band. Owing to the high transparency and conductivity of Ag-NWs/PMMA/PDMS FTEs, the phenomenon was achieved by in-situ PL measurement, as shown in Figure 6b. The PL intensity from the MoS_2 monolayer was significantly enhanced as the gate-voltage-induced electric field increased. The PL enhancement ratio was from 203.8% to 239.7% under the electric field intensity of 10.0–29.7 kV/cm, as demonstrated in Figure 6c, confirming the compatibility of Ag-NWs/PMMA/PDMS FTEs with 2D materials for electro-optical applications.

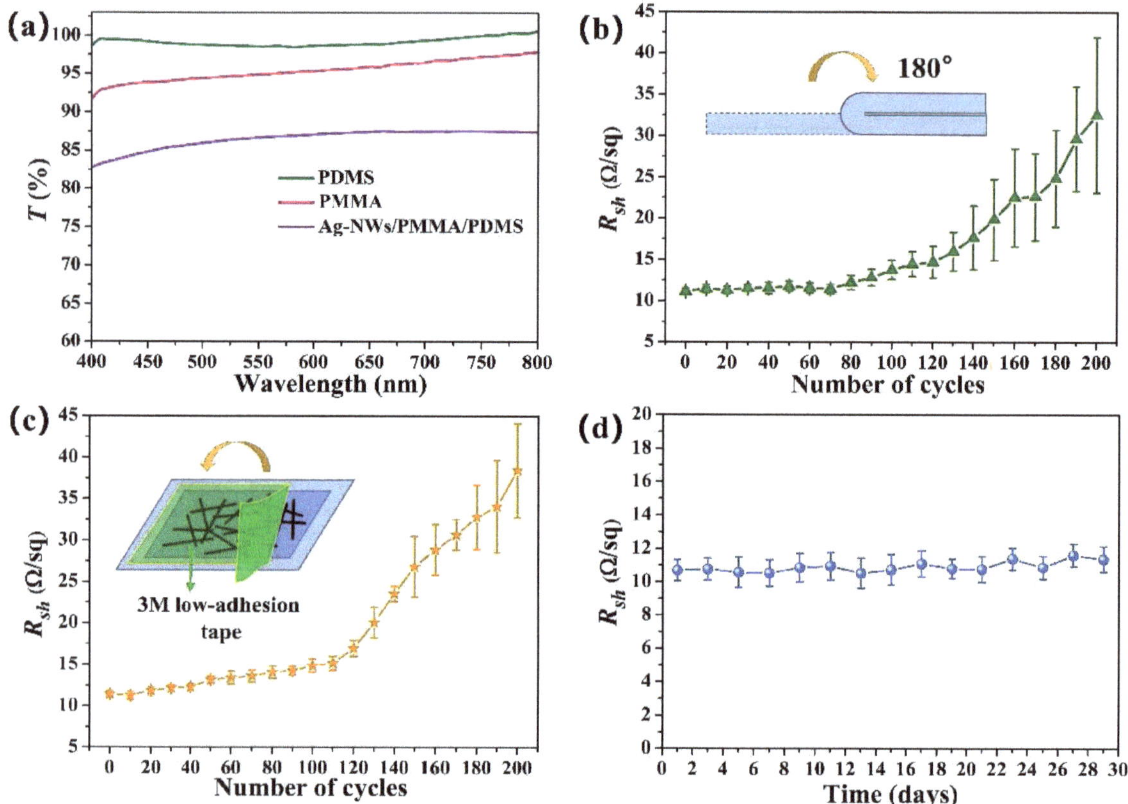

Figure 5. Durability of Ag-NWs/PMMA/PDMS FTEs. (**a**) Transmittance spectra in the visible band. (**b–d**) Durability tests for (**b**) flexibility, (**c**) adhesion, and (**d**) period stability.

The Ag-NWs/PMMA/PDMS films can also be used as facile, high-performance conductive tapes for conventional electronic components, where the good ohmic contacts with the footprints can be achieved by van der Waals and static electrical forces instead of by soldering. Figure 6d shows the *I–V* characteristics of a commercial surface-mounted LED with rigid metal electrodes and FTEs, respectively. It can be clearly seen that the ohmic contacts between FTEs and LED footprints were achieved. The increased opening voltage threshold of the LED from 1.7 V to 1.8 V was attributed to the inserted resistance of the FTEs. However, the *I–V* curve using FTEs was very similar to that of rigid metal electrodes. The photographs of luminescence from the LEDs with various electrodes under different bias voltages are shown Figure 6e. Although the inserted resistance slightly reduced the efficiency of the LED, the luminescence behaviors suggested the superiority of FTEs over conventional metal electrodes in future flexible electronic devices.

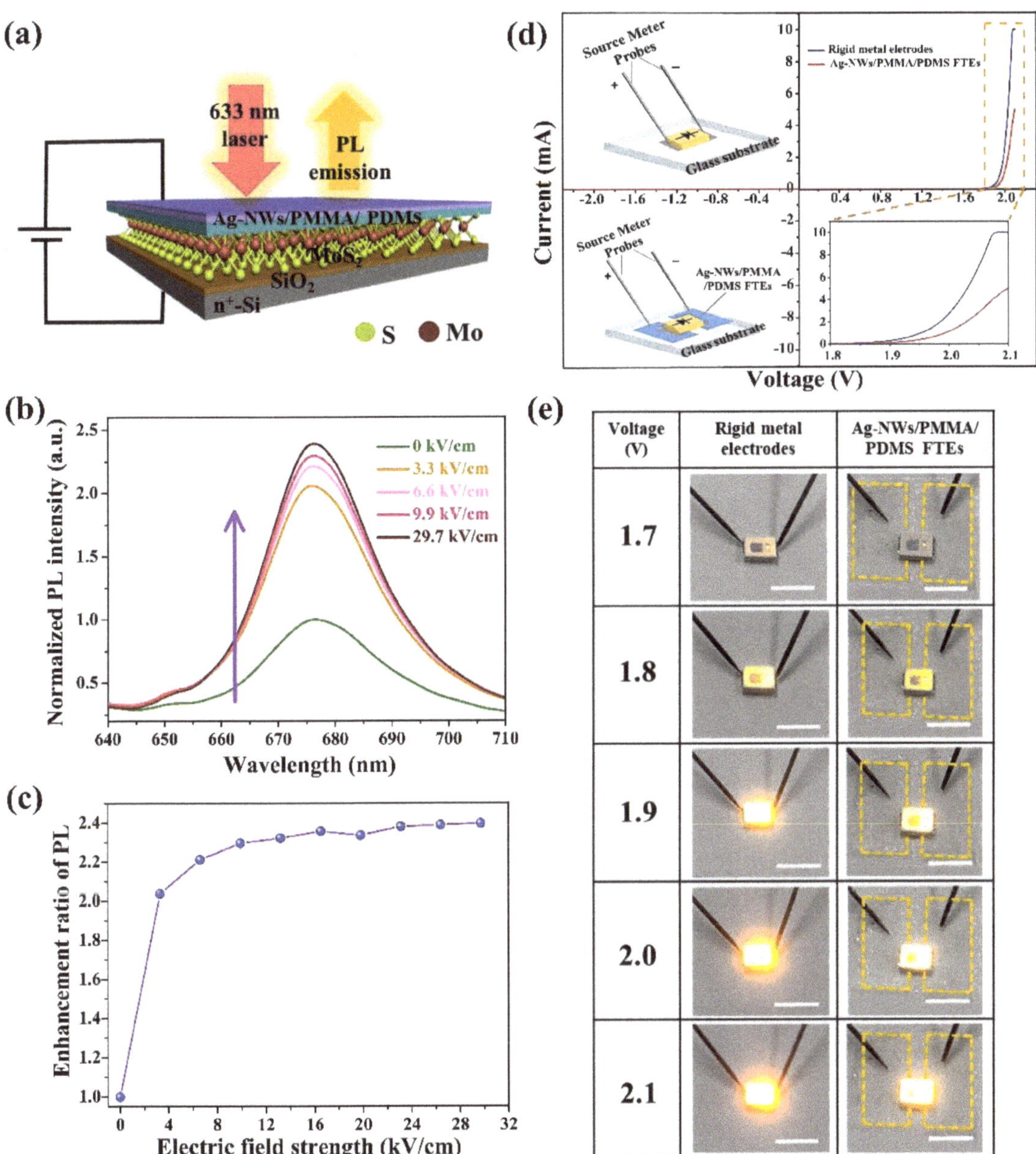

Figure 6. Applications of Ag-NWs/PMMA/PDMS FTEs. (**a**) Sandwich structure for gate-controlled PL emission from MoS_2 monolayer. (**b**) PL spectra of MoS_2 monolayer under various gate-voltage-induced electric field strengths. (**c**) Evaluation of PL enhancement ratio with electric field strength. (**d**) *I–V* curves of commercial LEDs with rigid metal electrodes and FTEs, where the insets demonstrate the experimental setups. (**e**) Photographs of luminescence from the LEDs under various bias voltages through different electrodes.

4. Conclusions

In this work, high-performance Ag-NWs/PMMA/PDMS FTEs were fabricated by the nanosecond pulsed laser nano-welding of Ag-NWs on PMMA/PDMS flexible substrates. Compared with previous fabrication techniques, laser nano-welding provided a facile and time-saving approach to improve Ag-NW network formation and adhesion onto the flexible organic substrates. The stimulated localized surface plasmon resonances confined the incident laser energy into the cross junctions and gaps between Ag-NWs for selective melting. The Ag-NWs were therefore welded for resistance reduction and jointed with the PMMA film with high adhesive strength. The fabricated FTEs demonstrated a high transmittance of 85.9% in the visible band and a low sheet resistance of 11.3 Ω/sq, attributable to the small absorption cross section and high conductivity of the nano-welded Ag-NWs. The superior durability, high flexibility, strong adhesion, and period stability tolerance commended the FTEs for practical wearable applications, in pursuit of which two typical components in 2D material luminescence devices and LEDs were explored. The gate voltages for high electric fields for regulating the PL emission from an MoS_2 monolayer was determined. The FTEs were also confirmed to be suitable for conventional electronic devices (i.e., LEDs), as a flexible-type electrode for wearable designs to replace rigid metal ones, since soldering between the FTEs and component footprints for ohmic contacts was avoided. The present work points a path forward in the development of high-performance FTEs for next-generation flexible electronic/optoelectronic devices with outstanding optical transmittance, high conductivity, and good durability.

Author Contributions: Conceptualization, Y.Y.; methodology, Y.Y. and T.W.; software, T.W. and J.H.; validation, T.W. and Y.Y.; formal analysis, T.W. and Y.Y.; investigation, T.W., Y.Y. and Y.W.; data curation, L.Z., Q.L., X.Z. (Xiaoxia Zhang), X.L., X.Z. (Xiaohua Zhang), and Y.P.; writing—original draft preparation, T.W.; writing—review and editing, Y.Y.; supervision, Y.Y. and Y.W.; project administration, Y.Y.; funding acquisition, Y.Y. All authors have read and agreed to the published version of the manuscript.

Funding: This research was funded by National Natural Science Foundation of China, grant number 12074019 and Key Program of Science and Technology Development Project of Beijing Municipal Education Commission, grant number KZ202110005002.

Data Availability Statement: The data presented in this study are available from the corresponding author.

Conflicts of Interest: The authors declare no conflict of interest.

References

1. Chen, W.; Liu, L.X.; Zhang, H.B.; Yu, Z.Z. Flexible, Transparent, and Conductive Ti_3C_2Tx MXene-Silver Nanowire Films with Smart Acoustic Sensitivity for High-Performance Electromagnetic Interference Shielding. *ACS Nano* **2020**, *14*, 16643–16653. [CrossRef]
2. Chen, X.; Xu, G.; Zeng, G.; Gu, H.; Chen, H.; Xu, H.; Yao, H.; Li, Y.; Hou, J.; Li, Y. Realizing Ultrahigh Mechanical Flexibility and >15% Efficiency of Flexible Organic Solar Cells via a "Welding" Flexible Transparent Electrode. *Adv. Mater.* **2020**, *32*, 1908478. [CrossRef]
3. De Guzman, N.; Ramos, M., Jr.; Donnabelle Balela, M. Improvements in the electroless deposition of Ag nanowires in hot ethylene glycol for resistive touchscreen device. *Mater. Res. Bull.* **2018**, *106*, 446–454. [CrossRef]
4. Sang, S.; Liu, L.; Jian, A.; Duan, Q.; Ji, J.; Zhang, Q.; Zhang, W. Highly sensitive wearable strain sensor based on silver nanowires and nanoparticles. *Nanotechnology* **2018**, *29*, 255202.
5. Triambulo, R.E.; Kim, J.-H.; Park, J.-W. Highly flexible organic light-emitting diodes on patterned Ag nanowire network transparent electrodes. *Org. Electron.* **2019**, *71*, 220–226. [CrossRef]
6. Hwang, Y.; Hwang, Y.H.; Choi, K.W.; Lee, S.; Kim, S.; Park, S.J.; Ju, B.-K. Highly stabilized flexible transparent capacitive photodetector based on silver nanowire/graphene hybrid electrodes. *Sci. Rep.* **2021**, *11*, 10499. [CrossRef]
7. Ma, C.; Liu, H.; Teng, C.; Li, L.; Zhu, Y.; Yang, H.; Jiang, L. Wetting-Induced Fabrication of Graphene Hybrid with Conducting Polymers for High-Performance Flexible Transparent Electrodes. *ACS Appl. Mater. Interfaces* **2020**, *12*, 55372–55381. [CrossRef]
8. Park, I.J.; Kim, T.I.; Yoon, T.; Kang, S.; Cho, H.; Cho, N.S.; Lee, J.I.; Kim, T.S.; Choi, S.Y. Flexible and Transparent Graphene Electrode Architecture with Selective Defect Decoration for Organic Light-Emitting Diodes. *Adv. Funct. Mater.* **2018**, *28*, 1704435. [CrossRef]

9. Fang, Y.; Li, Y.; Wang, X.; Zhou, Z.; Zhang, K.; Zhou, J.; Hu, B. Cryo-Transferred Ultrathin and Stretchable Epidermal Electrodes. *Small* **2020**, *16*, 2000450. [CrossRef]
10. Li, D.; Wang, L.; Ji, W.; Wang, H.; Yue, X.; Sun, Q.; Li, L.; Zhang, C.; Liu, J.; Lu, G.; et al. Embedding Silver Nanowires into a Hydroxypropyl Methyl Cellulose Film for Flexible Electrochromic Devices with High Electromechanical Stability. *ACS Appl. Mater. Interfaces* **2021**, *13*, 1735–1742. [CrossRef]
11. Wang, S.; Wu, S.; Ling, Z.; Chen, H.; Lian, H.; Portier, X.; Gourbilleau, F.; Marszalek, T.; Zhu, F.; Wei, B.; et al. Mechanically and thermally stable, transparent electrodes with silver nanowires encapsulated by atomic layer deposited aluminium oxide for organic optoelectronic devices. *Org. Electron.* **2020**, *78*, 105593. [CrossRef]
12. Zhang, Y.; Bai, S.; Chen, T.; Yang, H.; Guo, X. Facile preparation of flexible and highly stable graphene oxide-silver nanowire hybrid transparent conductive electrode. *Mater. Res. Express* **2020**, *7*, 016413. [CrossRef]
13. Che, B.; Zhou, D.; Li, H.; He, C.; Liu, E.; Lu, X. A highly bendable transparent electrode for organic electrochromic devices. *Org. Electron.* **2019**, *66*, 86–93. [CrossRef]
14. Jiang, S.; Hou, P.-X.; Chen, M.-L.; Wang, B.-W.; Sun, D.-M.; Tang, D.-M.; Jin, Q.; Guo, Q.-X.; Zhang, D.-D.; Du, J.-H.; et al. Ultrahigh-performance transparent conductive films of carbon-welded isolated single-wall carbon nanotubes. *Sci. Adv.* **2018**, *4*, eaap9264. [CrossRef]
15. Khachatryan, H.; Kim, K.-B.; Kim, M. Fabrication of Flexible Carbon Nanotube Network on Paper Substrate: Effect of Post Treatment Temperature on Electrodes. *J. Nanoelectron. Optoelecton.* **2020**, *15*, 1442–1449. [CrossRef]
16. Li, X.; Yu, S.; Zhao, L.; Wu, M.; Dong, H. Hybrid PEDOT:PSS to obtain high-performance Ag NW-based flexible transparent electrodes for transparent heaters. *J. Mater. Sci. Mater. Electron.* **2020**, *31*, 8106–8115. [CrossRef]
17. Song, J.; Ma, G.; Qin, F.; Hu, L.; Luo, B.; Liu, T.; Yin, X.; Su, Z.; Zeng, Z.; Jiang, Y.; et al. High-Conductivity, Flexible and Transparent PEDOT:PSS Electrodes for High Performance Semi-Transparent Supercapacitors. *Polymers* **2020**, *12*, 450. [CrossRef]
18. Yang, H.; Bai, S.; Chen, T.; Zhang, Y.; Wang, H.; Guo, X. Facile fabrication of large-scale silver nanowire-PEDOT:PSS composite flexible transparent electrodes for flexible touch panels. *Mater. Res. Express* **2019**, *6*, 086315. [CrossRef]
19. Tan, D.; Jiang, C.; Li, Q.; Bi, S.; Song, J. Silver nanowire networks with preparations and applications: A review. *J. Mater. Sci. Mater. Electron.* **2020**, *31*, 15669–15696. [CrossRef]
20. Park, J.; Hyun, B.G.; An, B.W.; Im, H.-G.; Park, Y.-G.; Jang, J.; Park, J.-U.; Bae, B.-S. Flexible Transparent Conductive Films with High Performance and Reliability Using Hybrid Structures of Continuous Metal Nanofiber Networks for Flexible Optoelectronics. *ACS Appl. Mater. Interfaces* **2017**, *9*, 20299–20305. [CrossRef] [PubMed]
21. Kim, H.; Choi, S.-H.; Kim, M.; Park, J.-U.; Bae, J.; Park, J. Seed-mediated synthesis of ultra-long copper nanowires and their application as transparent conducting electrodes. *Appl. Surf. Sci.* **2017**, *422*, 731–737. [CrossRef]
22. Ku, M.; Kim, J.; Won, J.-E.; Kang, W.; Park, Y.-G.; Park, J.; Lee, J.-H.; Cheon, J.; Lee, H.H.; Park, J.-U. Smart, soft contact lens for wireless immunosensing of cortisol. *Sci. Adv.* **2020**, *6*, eabb2891. [CrossRef]
23. Kim, J.; Park, J.; Park, Y.-G.; Cha, E.; Ku, M.; An, H.S.; Lee, K.-P.; Huh, M.-I.; Kim, J.; Kim, T.-S.; et al. A soft and transparent contact lens for the wireless quantitative monitoring of intraocular pressure. *Nat. Biomed. Eng.* **2021**, *5*, 772–782. [CrossRef]
24. Song, R.; Yao, S.; Liu, Y.; Wang, H.; Dong, J.; Zhu, Y.; O'Connor, B.T. Facile Approach to Fabricating Stretchable Organic Transistors with Laser-Patterned Ag Nanowire Electrodes. *ACS Appl. Mater. Interfaces* **2020**, *12*, 50675–50683. [CrossRef]
25. Qiu, J.; Wang, X.; Ma, Y.; Yu, Z.; Li, T. Stretchable Transparent Conductive Films Based on Ag Nanowires for Flexible Circuits and Tension Sensors. *ACS Appl. Nano Mater.* **2021**, *4*, 3760–3766. [CrossRef]
26. Giusti, G.; Langley, D.P.; Lagrange, M.; Collins, R.; Jimenez, C.; Brechet, Y.; Bellet, D. Thermal annealing effects on silver nanowire networks. *Int. J. Nanotechnol.* **2014**, *11*, 785–795. [CrossRef]
27. Oh, J.S.; Oh, J.S.; Yeom, G.Y. Invisible Silver Nanomesh Skin Electrode via Mechanical Press Welding. *Nanomaterials* **2020**, *10*, 633. [CrossRef] [PubMed]
28. Liu, Y.; Zhang, J.; Gao, H.; Wang, Y.; Liu, Q.; Huang, S.; Guo, C.F.; Ren, Z. Capillary-Force-Induced Cold Welding in Silver-Nanowire-Based Flexible Transparent Electrodes. *Nano Lett.* **2017**, *17*, 1090–1096. [CrossRef]
29. Yan, X.; Li, X.; Zhou, L.; Chu, X.; Yang, F.; Chi, Y.; Yang, X. Electrically sintered silver nanowire networks for use as transparent electrodes and heaters. *Mater. Res. Express* **2019**, *6*, 116316. [CrossRef]
30. Hong, C.H.; Oh, S.K.; Kim, T.K.; Cha, Y.J.; Kwak, J.S.; Shin, J.H.; Ju, B.K.; Cheong, W.S. Electron beam irradiated silver nanowires for a highly transparent heater. *Sci. Rep.* **2015**, *5*, 17716. [CrossRef]
31. Liang, X.; Lu, J.; Zhao, T.; Yu, X.; Jiang, Q.; Hu, Y.; Zhu, P.; Sun, R.; Wong, C.-P. Facile and Efficient Welding of Silver Nanowires Based on UVA-Induced Nanoscale Photothermal Process for Roll-to-Roll Manufacturing of High-Performance Transparent Conducting Films. *Adv. Mater. Interfaces* **2019**, *6*, 1801635. [CrossRef]
32. Hu, Y.; Liang, C.; Sun, X.; Zheng, J.; Duan, J.A.; Zhuang, X. Enhancement of the Conductivity and Uniformity of Silver Nanowire Flexible Transparent Conductive Films by Femtosecond Laser-Induced Nanowelding. *Nanomaterials* **2019**, *9*, 673. [CrossRef] [PubMed]
33. Lee, P.; Kwon, J.; Lee, J.; Lee, H.; Suh, Y.D.; Hong, S.; Yeo, J. Rapid and Effective Electrical Conductivity Improvement of the Ag NW-Based Conductor by Using the Laser-Induced Nano-Welding Process. *Micromachines* **2017**, *8*, 164. [CrossRef]
34. Liu, L.; Peng, P.; Hu, A.; Zou, G.; Duley, W.W.; Zhou, Y.N. Highly localized heat generation by femtosecond laser induced plasmon excitation in Ag nanowires. *Appl. Phys. Lett.* **2013**, *102*, 073107. [CrossRef]

35. Nian, Q.; Saei, M.; Xu, Y.; Sabyasachi, G.; Deng, B.; Chen, Y.P.; Cheng, G.J. Crystalline Nanojoining Silver Nanowire Percolated Networks on Flexible Substrate. *ACS Nano* **2015**, *9*, 10018–10031. [CrossRef]
36. Bell, A.R.; Fairfield, J.A.; McCarthy, E.K.; Mills, S.; Boland, J.J.; Baffou, G.; McCloskey, D. Quantitative Study of the Photothermal Properties of Metallic Nanowire Networks. *ACS Nano* **2015**, *9*, 5551–5558. [CrossRef]
37. Kang, T.; Yoon, I.; Jeon, K.-S.; Choi, W.; Lee, Y.; Seo, K.; Yoo, Y.; Park, Q.H.; Ihee, H.; Suh, Y.D.; et al. Creating Well-Defined Hot Spots for Surface-Enhanced Raman Scattering by Single-Crystalline Noble Metal Nanowire Pairs. *J. Phys. Chem. C* **2009**, *113*, 7492–7496. [CrossRef]
38. Prokes, S.M.; Alexson, D.A.; Glembocki, O.J.; Park, H.D.; Rendell, R.W. Effect of crossing geometry on the plasmonic behavior of dielectric core/metal sheath nanowires. *Appl. Phys. Lett.* **2009**, *94*, 093105. [CrossRef]
39. Lei, D.Y.; Aubry, A.; Maier, S.A.; Pendry, J.B. Broadband nano-focusing of light using kissing nanowires. *New J. Phys.* **2010**, *12*, 093030. [CrossRef]
40. Haacke, G. New figure of merit for transparent conductors. *J. Appl. Phys.* **1976**, *47*, 4086–4089. [CrossRef]
41. Park, Y.-G.; Kim, H.; Park, S.-Y.; Kim, J.-Y.; Park, J.-U. Instantaneous and Repeatable Self-Healing of Fully Metallic Electrodes at Ambient Conditions. *ACS Appl. Mater. Interfaces* **2019**, *11*, 41497–41505. [CrossRef]
42. Samy, O.; Zeng, S.; Birowosuto, M.D.; El Moutaouakil, A. A Review on MoS_2 Properties, Synthesis, Sensing Applications and Challenges. *Crystals* **2021**, *11*, 355. [CrossRef]
43. Newaz, A.K.M.; Prasai, D.; Ziegler, J.I.; Caudel, D.; Robinson, S.; Haglund, R.F., Jr.; Bolotin, K.I. Electrical control of optical properties of monolayer MoS_2. *Solid State Commun.* **2013**, *155*, 49–52. [CrossRef]
44. Yi, H.T.; Rangan, S.; Tang, B.; Frisbie, C.D.; Bartynski, R.A.; Gartstein, Y.N.; Podzorov, V. Electric-field effect on photoluminescence of lead-halide perovskites. *Mater. Today* **2019**, *28*, 31–39. [CrossRef]

MDPI

Article

Mobility of Small Molecules in Solid Polymer Film for π-Stacked Crystallization

Yue Liu and Xinping Zhang *

Institute of Information Photonics Technology, Beijing University of Technology, Beijing 100124, China; liuyue2020@emails.bjut.edu.cn
* Correspondence: zhangxinping@bjut.edu.cn

Abstract: Crystallization or π-stacked aggregation of small molecules is an extensively observed phenomenon which favors charge transport along the crystal axis and is important for the design of organic optoelectronic devices. Such a process has been reported for *N,N′*-Bis(1-ethylpropyl)-3,4,9,10-perylenebis(dicarboximide) (EPPTC). However, the π-stacking mechanism requires solution–air or solution–solid interfaces. The crystallization or aggregation of molecules doped in solid films is generally thought to be impossible, since the solid environment surrounding the small molecules does not allow them to aggregate together into π-stacked crystals. In this work, we demonstrate that the movement of the EPPTC molecules becomes possible in a solid polymer film when it is heated to above the glass transition temperature of the polymer. Thus, crystal particles can be produced as a doped matrix in a thin solid film. The crystallization process is found to be strongly dependent on the annealing temperature and the annealing time. Both the microscopic and spectroscopic evaluations verify such discoveries and characterize the related properties of these crystals.

Keywords: small molecules; crystallization; moving in solid film; glass transition temperature; annealing temperature; annealing time

Citation: Liu, Y.; Zhang, X. Mobility of Small Molecules in Solid Polymer Film for π-Stacked Crystallization. *Crystals* **2021**, *11*, 1022. https://doi.org/10.3390/cryst11091022

Academic Editor: Emilio Parisini

Received: 1 August 2021
Accepted: 23 August 2021
Published: 25 August 2021

1. Introduction

Crystallization of organic molecules through π-stacking is an important approach to improve the charge transport performance in semiconductors and achieve high-efficiency optoelectronic devices [1–3]. Different forms of crystalline materials or structures have been reported to produce organic semiconductors with much-improved optical and electronic properties [4–7]. Various techniques have been employed to produce organic molecular structures with oriented aggregation and ordered distribution [8,9]. Thermal annealing is a convenient and efficient approach to realizing such molecular rearrangement or surface modification processes [10–16]. The precondition for such crystalline aggregation is the planar molecular structure, and this can generally be satisfied in small molecules, which allow face-to-face aggregation between one another.

Perylene and its derivatives have high electron mobility, which makes them promising candidates for organic photovoltaic and transistor devices [6,7,17,18]. Due to the planar structures of the perylene molecules, they can easily aggregate into one-dimensional crystals at the solution–solid, vapor–solid, and solution–air interfaces, as well as at those between different solvents [1,3,17,19]. However, such a crystallization process has not been observed within solids. In this work, we report the crystallization of perylene molecules in the solid thin film of polymers, where the polymer needs to be annealed to its glass transition temperature so that the solid film becomes sufficiently softened and flexibilized to allow the motion of the doped small molecules. The corresponding mechanisms are revealed not only by the microscopic and spectroscopic properties, but also by the annealing temperature- and annealing time-dependence of the crystallization process. Blending small molecules with polymers is an important approach to construct efficient optoelectronic devices based on heterojunctions. New charge transfer complexes [20–22] can be thus produced and utilized to investigate new photophysics and to develop new photovoltaic devices.

Figure 1 shows the basic principles for the formation of the EPPTC crystals inside a solid film of polymer chains. In Figure 1a, we illustrate a matrix of small molecules of perylene (*N*,*N*′-Bis(1-ethylpropyl)-3,4,9,10-perylenebis(dicarboximide) (EPPTC)) doped in a solid film, which are distributed randomly without any orientational arrangement. In our approach, we used polymethyl methacrylate (PMMA) to supply the polymeric solid film. Owing to the planar molecular shape, π-stacking between the EPPTC molecules becomes possible, provided that channels are supplied for the EPPTC molecules to move across the PMMA molecular chains.

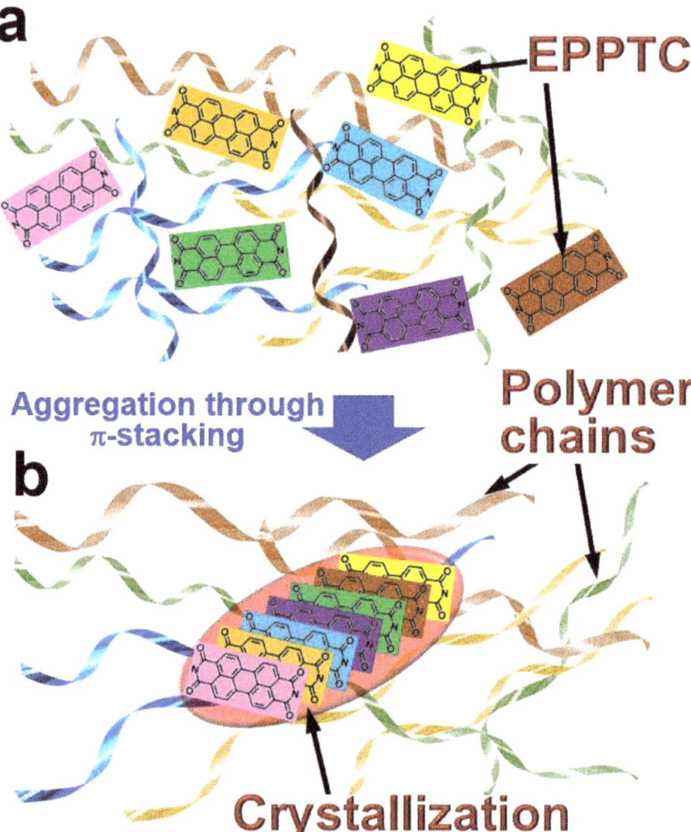

Figure 1. (**a**) Distribution of the EPPTC molecules in the solid film of PMMA molecular chains. (**b**) Crystallization of EPPTC small molecules due to their motion across the annealing-flexibilized PMMA molecular chains.

The crossing channel can be produced by heating the solid polymer film to the glass transition temperature, which is 100–120 °C. Therefore, 120 °C was measured as the threshold temperature for the crystallization process of EPPTC in PMMA film. The softened PMMA molecular chains allow the small molecules to move across the solid film and become aggregated, forming π-stacked crystals after the cooling down of the whole sample, as shown in Figure 1b. Apparently, the crystals are roughly in fusiform shapes, which are distributed randomly in their orientations and sizes. It is also understood that the length and thickness of the crystals are dependent on the annealing temperature and annealing time. Furthermore, since the glass transition has a threshold temperature, the formation of the crystal particles in the solid polymer film also has a threshold temperature, which is found to be about 120 °C. Below this temperature, the crystallization process illustrated in Figure 1b is found to not be possible even with a long time of annealing.

2. Annealing-Aided Crystallization of Small Molecules in Polymeric Solids

2.1. Annealing Temperature Dependence

In the experiments, we first prepared solutions of EPPTC and PMMA in chloroform with a concentration of 10 mg/mL and 15 mg/mL, respectively. A mixture solution was prepared by mixing the above two solutions with a volume-to-volume ratio of 1:1. A blend film was then prepared by spin-coating the mixture solution onto a glass substrate with a speed of 2000 rpm for 30 s. Multiple samples were prepared under the same conditions so that comparison could be made between different annealing processes.

For the annealing, we employed five temperatures: 30, 60, 90, 120, and 150 °C. However, for annealing temperatures below 120 °C, nearly no changes were observed either in the microscopic or in the spectroscopic response of the sample. This is clearly because the glass transition temperature was not reached. In Figure 2, we present only the scanning electron microscope (SEM, JEOL, Tokyo, Japan) and atomic force microscope (AFM, Witec GmbH, Ulm, Germany) images of the samples annealed at 120 and 150 °C. The SEM and AFM images of the sample annealed at 120 °C are shown in Figure 2a,b, respectively. Since the annealing temperature is already above the threshold for glass transition, crystal particles with very small sizes and very low densities can be observed in both images. The mean diameter of the crystal particles is roughly 140 nm. We need to note that the annealing time was 10 min and the size and density of the crystal particles may be increased by extending the annealing time. Thus, it is verified that the glass transition temperature is the threshold for the crystallization of EPPTC molecules, where the softened PMMA molecular chains allow the movement of the EPPTC molecules and their aggregation through π-stacking.

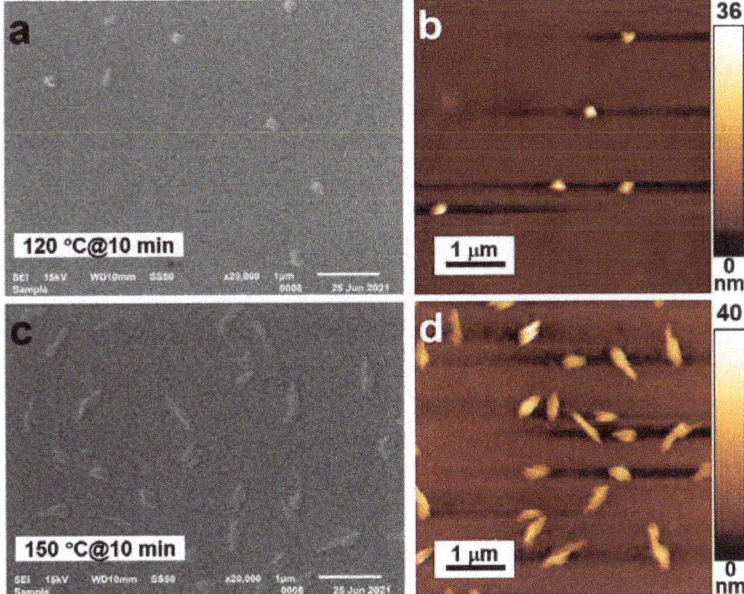

Figure 2. SEM (**a**,**c**) and AFM (**b**,**d**) images for the samples annealed at 120 (**a**,**b**) and 150 °C (**c**,**d**).

When the annealing temperature was increased to 150 °C, for the same annealing time of 10 min, the crystal length and thickness were increased dramatically, as shown by the SEM and AFM images in Figure 2c,d, respectively. Using a rough evaluation, the crystals have a mean length of about 900 nm and a width of 160 nm. Meanwhile, the height of the crystals was also increased slightly from about 36 to 40 nm. Furthermore, the density of the crystal particles was more obviously increased. According to Figure 2c,d, the number of crystal particles was increased by a factor of five, corresponding to an increase in the total number of crystal particles from about 6 to 30. This implies convincingly that above the glass transition temperature the PMMA molecules become more flexible and allow the strong aggregation of the EPPTC molecules into crystals.

It is also understood that the crystal particles formed by molecular aggregation through π-stacking greatly influence the spectroscopic properties of the small molecules. Figure 3a shows the absorption spectra at different annealing temperatures, and Figure 3b,c show the normalized data in Figure 3a and the PL spectra of the blend film, respectively, for different annealing temperatures. As the temperature was increased from 30 to 120 °C, where the annealing time was 10 min, a weak feature appeared at about 474 nm and the relative intensity at 530 nm was reduced, as shown in Figure 3b. However, when the temperature was increased to 150 °C, which exceeds the glass transition temperature, the spectral feature became obvious at 473 nm (dashed upward arrow in black) and the peak at about 530 nm suddenly became a valley in the absorption spectrum (dashed downward arrow), as shown in Figure 3b. Furthermore, the absorption spectrum peak at about 542 nm appeared in a sudden manner (dashed upward arrow in red). All of these changes highlighted by dashed arrows resulted from the aggregation of the EPPTC molecules in the softened PMMA film, or from the crystallization through π-stacking of the EPPTC molecules.

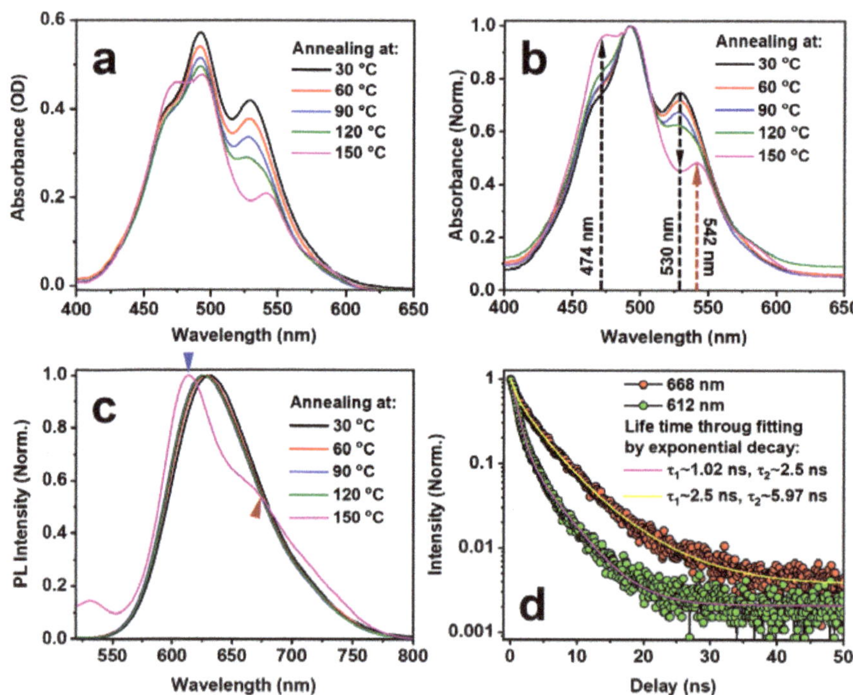

Figure 3. (**a**) Absorption spectra measured at different annealing temperatures. (**b**) Normalized absorption spectra in (**a**). (**c**) Normalized PL spectra measured at different annealing temperatures. (**d**) Fluorescence lifetime at 612 (blue triangle in (**c**)) and 668 nm (red triangle in (**c**)) for the sample annealing at 150 °C for 10 min.

Figure 3c shows the PL spectra annealed at 30, 60, 90, 120, and 150 °C for 10 min, where all spectra have been normalized. Obviously, with an annealing temperature lower than 120 °C, there is little change in the PL spectra. However, when the annealing temperature exceeds 150 °C, the PL spectrum changed dramatically: the main peak shifts from about 630 nm to 612 nm (blue triangle) and a new peak appears at about 668 nm (red triangle). According to our previous work [17], the emission at 612 nm corresponds mainly to the intrinsic molecular emission from EPPTC and that at 668 nm mainly to the crystal phase emission. Figure 3d shows the measurements of the emission dynamics at 612 and 668 nm using green and red circles, respectively. At 612 nm, the emission lifetime is in the range of 1~2.5 ns. However, the lifetime extends to about 6 ns at 668 nm, which can be identified as the lifetime of the emission from EPPTC crystal phase. Therefore, the spectroscopic response confirms the crystallization of the EPPTC molecules doped in the PMMA solid film when the sample is annealed at a temperature above the glass transition threshold. It needs to be noted that since the glass transition temperature is different for different polymers and for polymers with different molecular weights, the annealing strategy should be modified for different polymer materials.

2.2. Annealing Time Dependence

At a temperature above the glass transition threshold of PMMA, it takes time for the EPPTC molecules to become aggregated into crystal particles. Figure 4 shows the SEM images of the sample annealed at 150 °C for different times. Before annealing, no crystal particles can be observed in Figure 4a, implying that no crystallization process took place before annealing. Crystal particles can be resolved in Figure 4b for an annealing time of 30 s, although they are small and appear with a low density. For an annealing time longer than 1 min, EPPTC crystal particles can be observed clearly on the surface of the PMMA film. Figure 4c–f show the SEM images for annealing times of 1, 2, 5, and 10 min, respectively. Obviously, both the size and density of the EPPTC crystal particles increase dramatically with increased annealing time. The crystal particles can be clearly resolved only for an annealing time longer than 2 min, and the morphology of the surface of the solid film remains nearly constant when the sample is annealed for more than 5 min. The mean width and length of the crystals are roughly the same as those demonstrated in Figure 2c,d. It is also clearly observable in Figure 4 that the crystals are distributed randomly with random orientations, and that each particle contains inhomogeneous structures, implying that the crystal particles are not single crystals.

Figure 5 shows the spectroscopic characterization of the EPPTC-doped PMMA thin film sample, which was annealed at 150 °C for different times. Figure 5a,b show the directly measured and normalized absorption spectra, respectively. The main absorption peak at about 490 nm did not change its spectral position during the whole process. At least four other new features can be observed with increased annealing time, as highlighted by the dashed arrows in Figure 5b. The absorption peak around 473 nm was enhanced with increased annealing time, while that at about 530 nm is reduced and a new peak appeared at 542 nm; these features are similarly observed in Figure 3b. However, the most obvious difference between Figures 3b and 5b is the spectral peak at about 578 nm, which was observed clearly when the annealing time was longer than 1 min. Therefore, this is the more typical absorption spectrum of the crystal phase.

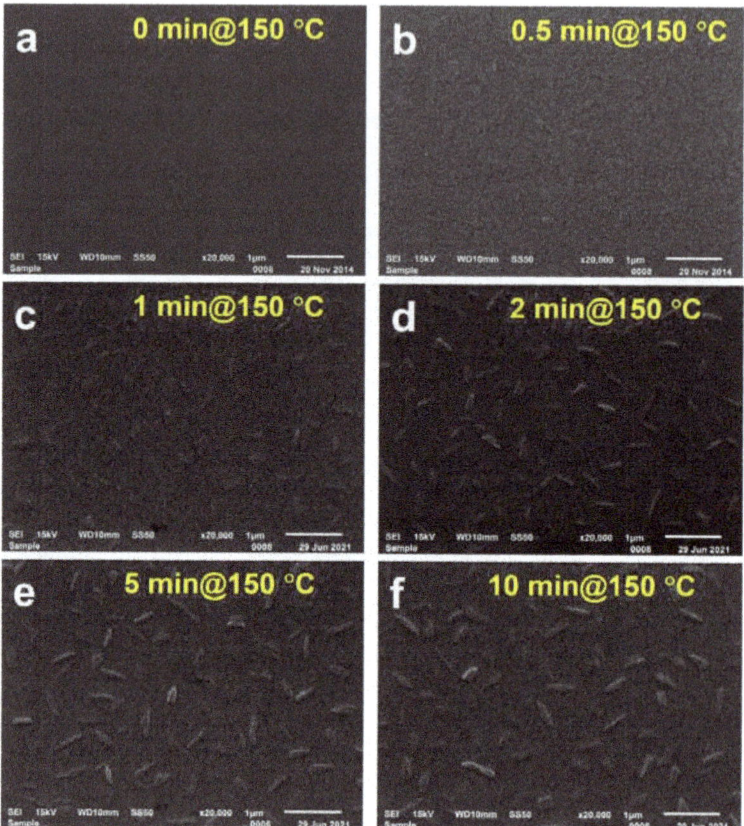

Figure 4. SEM images at an annealing temperature of 150 °C for (**a**) 0, (**b**), 0.5, (**c**) 1, (**d**) 2, (**e**) 5, and (**f**) 10 min.

As has been discussed with regard to Figure 3c, the PL spectrum becomes separated into two components for an annealing temperature of 150 °C. This is again observed in Figure 5c where the annealing time is longer than 1 min, which agrees with the observation of the time-dependent absorption in Figure 5b. The spectral feature peaked at a longer wavelength of 668 nm; this resulted from the crystal phase emission, as highlighted by the dashed arrow in Figure 5c. The difference between the crystal phase emission and the intrinsic molecular emission lies not only in their spectral positions but also in their lifetimes. This is shown in Figure 5d, where we demonstrate the PL dynamics at 612 and 668 nm for different annealing times. Clearly, the emission from the crystal phase at 668 nm is much longer than that from the intrinsic emission from EPPTC molecules.

From Figure 5d, we can also observe that the longest emission lifetime corresponds to an annealing time of 30 s at 150 °C, and that emission lifetime reduces with increased annealing time. For an annealing time longer than 1 minute, the decay dynamics of the emission remain nearly constant. This implies that there exists an intermediate state for the formation of the crystal phase with less quenching of the photoluminescence due to intermolecular interactions. However, after 10 min of annealing, the intrinsic EPPTC emission is much quenched due to the intermolecular interaction in the crystal phase; therefore, emission lifetimes at 612 and 668 nm are both reduced. The lifetime values become fixed after the equilibrium of the crystallization process is reached at a specified annealing temperature for a sufficiently long annealing time. We can also identify clearly from Figure 5d that the emission lifetime at 668 nm is much longer than that at

612 nm, implying long-lived crystal phase emissions, as compared with the intrinsic EPPTC molecular emissions, verifying the formation of the π-stacked crystals.

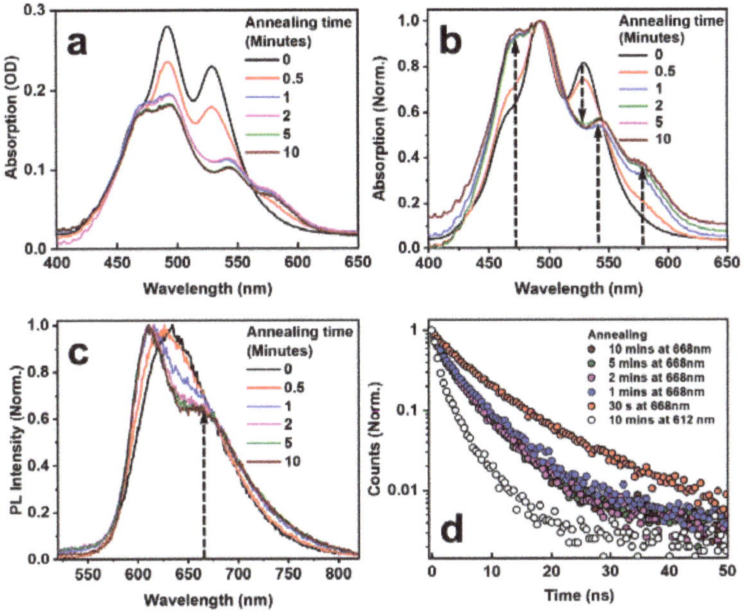

Figure 5. Absorption (**a**,**b**) and PL (**c**) spectra of the blend film annealed at 150 °C for different annealing times. (**b**) A normalized replot of (**a**). (**d**) Emission decay dynamics at 668 nm for different annealing times of 0.5, 1, 2, 5, 10 min and a comparison with that at 612 nm for an annealing time of 10 min.

3. Conclusions

We have demonstrated in this experimental work that it is possible for small molecular semiconductors to move and aggregate into crystal phases in a solid film of polymers, as long as the polymer film is softened and flexibilized, e.g., through heating it to its glass transition temperature. We verified such mechanisms using EPPTC molecules doped in a PMMA thin film. Both the microscopic and spectroscopic evaluations justified the validness of the proposed mechanisms and characterized the properties of the crystalized materials/structures. We can thus conclude that heterojunctions based on blends of small molecules and polymers can be constructed and modified by a low-temperature annealing process after production of the solid film, which simplifies the preparation and enables high-quality thin-film devices to be produced. Therefore, this strategy is important for the design and realization of organic optoelectronic devices based on blend materials or heterojunction structures.

Author Contributions: Conceptualization, X.Z.; Formal analysis, X.Z.; Funding acquisition, X.Z.; Investigation, Y.L. and X.Z.; Methodology, X.Z.; Project administration, X.Z.; Resources, X.Z.; Supervision, X.Z.; Writing—original draft, X.Z.; Writing—review & editing, X.Z. All authors have read and agreed to the published version of the manuscript.

Funding: This research was funded by National Natural Science Foundation of China, grant number 61735002 and 12074020 and by Beijing Municipal Education Commission, grant number KZ202010005002. The APC was funded by National Science Foundation of China.

Institutional Review Board Statement: Not applicable.

Informed Consent Statement: Not applicable.

Data Availability Statement: Not applicable.

Acknowledgments: The authors acknowledge the National Natural Science Foundation of China (61735002, 12074020) and Beijing Municipal Education Commission (KZ202010005002) for their support.

Conflicts of Interest: The authors declare no conflict of interest.

References

1. Nguyen, T.-Q.; Martel, R.; Avouris, P.; Bushey, M.L.; Brus, L.; Nuckolls, C. Molecular Interaction in One-Dimensional Organic nanostructures. *J. Am. Chem. Soc.* **2004**, *126*, 5234. [CrossRef]
2. Karl, N.; Kraft, K.-H.; Marktanner, J.; Munch, M.; Schatz, F.; Stehle, R.; Uhde, H.-M. Fast electronic transport in organic molecular solids? *J. Vac. Sci. Technol. A* **1999**, *17*, 2318. [CrossRef]
3. Hill, J.P.; Jin, W.; Kosaka, A.; Fukushima, T.; Ichihara, H.; Shimomura, T.; Ito, K.; Hashizume, T.; Ishii, N.; Aida, T. Self-assembled hexa-peri-hexabenzocoronene graphitic nanotube. *Science* **2004**, *304*, 1481. [CrossRef]
4. Mas-Torrent, M.; Durkut, M.; Hadley, P.; Ribas, X.; Rovira, C. High Mobility of Dithiophene-Tetrathiafulvalene Single-Crystal Organic Field Effect Transistors. *J. Am. Chem. Soc.* **2004**, *126*, 984. [CrossRef]
5. de Boer, R.W.I.; Gershenson, M.E.; Morpurgo, A.F.; Podzorov, V. Organic single-crystal field-effect transistors. *Phys. Stat. Solidi A* **2004**, *201*, 1302. [CrossRef]
6. Xu, B.; Xiao, X.; Yang, X.; Zang, L.; Tao, N. Large gate modulation in the current of a room Temperature single molecule transistor. *J. Am. Chem. Soc.* **2005**, *127*, 2386. [CrossRef]
7. Schmidt-Mende, L.; Fechtenkoetter, A.; Muellen, K.; Moons, E.; Friend, R.H.; Mackenzie, J.D. Self-organized discotic liquid crystals for high-efficiency organic photovoltaics. *Science* **2001**, *293*, 1119. [CrossRef]
8. Ohisa, S.; Pu, Y.; Yamada, N.; Matsuba, G.; Kido, J. Influence of solution- and thermal-annealing processes on the sub-nanometer-ordered organic–organic interface structure of organic light-emitting devices. *Nat. Commun.* **2017**, *9*, 25–30. [CrossRef]
9. Wuerthner, F. Perylene bisimide dyes as versatile building blocks for functional supermolecular architechtures. *Chem. Commun.* **2004**, *2004*, 1564–1579. [CrossRef]
10. Lee, B.; Park, O. The effect of different heat treatments on the luminescence efficiency of polymer light-emitting diodes. *Adv. Mater.* **2000**, *12*, 801–804. [CrossRef]
11. Ahn, J.; Wang, C.; Widdowson, N.; Pearson, C.; Bryce, M.; Petty, M. Thermal annealing of blended-layer organic light-emitting diodes. *J. Appl. Phys.* **2005**, *98*, 054508. [CrossRef]
12. Sajjad, M.; Zhang, Y.; Geraghty, P.; Mitchell, V.; Ruseckas, A.; Blaszczyk, O.; Jones, D.; Samuel, I. Tailoring exciton diffusion and domain size in photovoltaic small molecules by annealing. *J. Mater. Chem. C* **2019**, *7*, 7922–7928. [CrossRef]
13. Tanaka, H.; Abe, Y.; Matsuo, Y.; Kawai, J.; Soga, I.; Sato, Y.; Nakamura, E. An amorphous mesophase generated by thermal annealing for high-performance organic photovoltaic devices. *Adv. Mater.* **2012**, *24*, 3521–3525. [CrossRef]
14. Jaqadamma, L.; Sajjad, M.; Savikhin, V.; Toney, M.; Samuel, I. Correlating photovoltaic properties of a PTB7-Th:PC_{71}BM blend to photophysics and microstructure as a function of thermal annealing. *J. Mater. Chem. A* **2017**, *5*, 14646–14657. [CrossRef]
15. Zhang, T.; Han, H.; Zou, Y.; Lee, Y.; Oshima, H.; Wong, K.; Holmes, R. Impact of thermal annealing on organic photovoltaic cells using regioisomeric donor-acceptor-acceptor molecules. *ACS Appl. Mater. Interfaces* **2017**, *9*, 25418–25425. [CrossRef]
16. Yan, B.; Swaraj, S.; Wang, C.; Hwang, I.; Greenham, N.; Groves, C.; Ade, H.; Mcneill, C. Influence of annealing and interfacial roughness on the performance of bilayer donor/acceptor polymer photovoltaic devices. *Adv. Funct. Mater.* **2010**, *20*, 4329–4337. [CrossRef]
17. Zhang, X.P.; Sun, B.Q. Organic crystal fibers aligned into oriented bundles with polarized emission. *J. Phys. Chem. B* **2007**, *111*, 10881–10885. [CrossRef]
18. van Herrikhuyzen, J.; Syamakumari, A.; Schenning, A.P.H.J.; Meijer, E.W. Synthesis of n-type perylene bisimide derivatives and their orthogonal self-assembly with p-type oligo(p-phenylene vinylene)s. *J. Am. Chem. Soc.* **2004**, *126*, 10021. [CrossRef]
19. Mascaro, D.J.; Thompson, M.E.; Smith, H.I.; Bulovic, V. Forming oriented organic crystals from amorphous thin films on patterned substrates via solvent-vapor annealing. *Org. Electron.* **2005**, *6*, 211. [CrossRef]
20. Zhang, X.P.; Dou, F.; Liu, H.M. Charge-Transfer Complex Coupled Between Polymer and H-Aggregate Molecular Crystals. *J. Polym. Sci. B Polym. Phys.* **2013**, *51*, 749–755. [CrossRef]
21. Garner, L.E.; Viswanathan, V.N.; Arias, D.H.; Brook, C.P.; Christensen, S.T.; Ferguson, A.J.; Kopidakis, N.; Larson, B.W.; Owczarczyk, Z.R.; Pfeilsticker, J.R.; et al. Photobleaching dynamics in small molecule vs. polymer organic photovoltaic blends with 1,7-bis-trifluoromethylfullerene. *J. Mater. Chem. A* **2018**, *6*, 4623. [CrossRef]
22. Li, H.; Lu, K.; Wei, Z.X. Polymer/small molecule/fullerene based ternary solar cells. *Adv. Energy Mater.* **2017**, *7*, 1602540. [CrossRef]

Article

Two-Dimensional Crystalline Gridding Networks of Hybrid Halide Perovskite for Random Lasing

Jingyun Hu [1], Haibin Xue [2] and Xinping Zhang [1,*]

1 Institute of Information Photonics Technology, Beijing University of Technology, Beijing 100124, China; hujingyun@emails.bjut.edu.cn
2 Eighth Medical Center of the General Hospital of the Chinese People's Liberation Army, Beijing 100091, China; tornatorex@163.com
* Correspondence: zhangxinping@bjut.edu.cn

Abstract: We report fabrication of large-scale homogeneous crystallization of $CH_3NH_3PbBr_3$ ($MAPbBr_3$) in the patterned substrate by a two-dimensional (2D) grating. This achieves high-quality optoelectronic structures on local sites in the micron scales and a homogeneous thin-film device in a centimeter scale, proposing a convenient technique to overcome the challenge for producing large-area thin-film devices with high quality by spin-coating. Through matching the concentration of the $MAPbBr_3$/DMF solutions with the periods of the patterning structures, we found an optimized size of the patterning channels for a specified solution concentration (e.g., channel width of 5 µm for a concentration of 0.14 mg/mL). Such a design is also an excellent scheme for random lasing, since the crystalline periodic networks of $MAPbBr_3$ grids are multi-crystalline constructions, and supply strong light-scattering interfaces. Using the random lasing performance, we can also justify the crystallization qualities and reveal the responsible mechanisms. This is important for the design of large-scale optoelectronic devices based on thin-film hybrid halide perovskites.

Keywords: two-dimensional patterning; periodical grids; hybrid halide perovskites; random lasing; large-scale thin-film networks

Citation: Hu, J.; Xue, H.; Zhang, X. Two-Dimensional Crystalline Gridding Networks of Hybrid Halide Perovskite for Random Lasing. *Crystals* **2021**, *11*, 1114. https://doi.org/10.3390/cryst11091114

Academic Editor: Luisa De Marco

Received: 17 August 2021
Accepted: 11 September 2021
Published: 13 September 2021

Publisher's Note: MDPI stays neutral with regard to jurisdictional claims in published maps and institutional affiliations.

1. Introduction

Organic-inorganic hybrid halide perovskites are a group of promising semiconductor materials for high-efficiency optoelectronic devices [1–6]. Light-emitting [7–9] and photovoltaic [10–12] diodes, as well as lasing devices [13–17], have been reported on extensively. Although single- or multi-crystals have been synthesized in large scales at high quality [18–22] and have been applied for various purposes, thin-film devices [23–26] are more attractive in the construction of devices that are integratable into micro- or nano-scale systems. However, the conventional spin-coating method may encounter problems with the homogeneity of the produced film, in particular for large-scale fabrication. This is not only because of the dewetting of the solution on the conventional substrates, but also due to the quick crystallization of the molecules in random scales and random distributions.

Patterning the substrate with designed micro- or nano-structures and optimized microscopic dimensions may not only modify its surface-energy properties, but also control the molecular crystallization process by dividing the large-area surface into periodic/nonperiodic localized sites. This is an ideal strategy to achieve large-area thin films with high qualities precisely controllable on each local site. Such a strategy applies not only to hybrid halide perovskites [27] but also to any other organic/inorganic semiconductors [28,29] or even biomolecules [30] with strong aggregation performance. The corresponding designs also simultaneously facilitate new functions, including micro-/nano-cavity effects, local-field confinement, output coupling control, optical waveguides, and interface structural optimization or metasurface incorporation [31]. We have recently reported controllable crystallization of the hybrid halide perovskites into grating lines with continuous distribution over a large length scale [27], which provides an effective approach to overcome the above-mentioned challenge for producing high-quality $MAPbBr_3$ crystal stripe gratings with controllable performance.

In this work, we report fabrication of large-area thin films into two dimensions, which enabled strong confinement of the molecules into more localized sites, and thus higher-quality crystallization. An optimized matching was found between the solution concentration and the patterning periods, with a duty cycle of about 50%. The random lasing performance was investigated to characterize the fabricated 2D structures and to verify the responsible mechanisms. Thus, we achieved thin-film $CH_3NH_3PbBr_3$ ($MAPbBr_3$) with a homogeneous area in the scale of cm^2 and locally high-quality crystalline structures in sizes of microns.

2. The Network of Squarely Gridded Thin Film of $MAPbBr_3$

The preparation method is illustrated in Figure 1 for the gridding network of the $MAPbBr_3$ thin film. First, a two-dimensional (2D) grating was produced by photolithography, where a positive photoresist S1805 was employed, so that the grating consists of periodically distributed square cylinders on a glass substrate, as shown in Figure 1a. Using different masks, we have fabricated gratings with different periods of 10, 20, and 50 μm. Then, the solution of $MAPbBr_3$ in DMF with different concentrations was spin-coated onto the top surface of the grating. In the last stage, the sample was heated to 90 °C for about 60 s. Figure 1c,d shows the microscopic characterization of the template photoresist grating by 3D profiler and the finalized $MAPbBr_3$ grating by a fluorescence optical microscope, respectively. They illustrate clearly high-quality 2D gratings of square cylinders with steep edges and homogeneous 2D grids of crystallized $MAPbBr_3$. Since Figure 1 is used to interpret the preparation procedures, we did not present detailed parameters of the structures in Figure 1c,d. We have also found that for a concentration of 0.14 mg/mL, the gridding network surrounding the cylinders can be filled completely and homogeneously, as illustrated in Figure 1b. Higher concentrations led to inhomogeneity and connection between neighboring locations, while concentrations that were too low led to incomplete fill of the grating grooves.

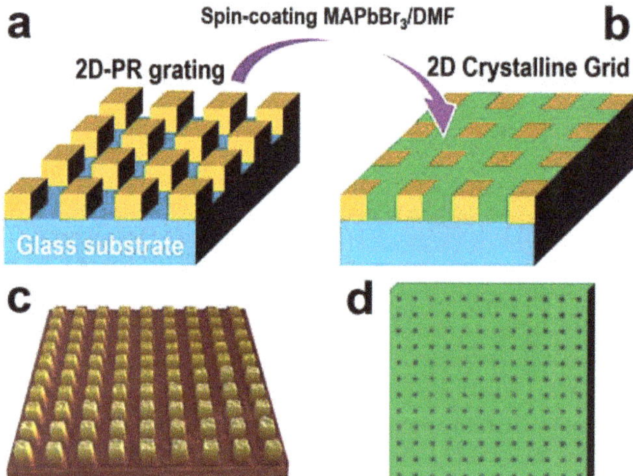

Figure 1. (**a**,**b**): Schematic illustration of the fabrication of gridded $MAPbBr_3$ crystalline film by filling the void space in the 2D PR grating through spin-coating the $MAPbBr_3$/DMF solution. (**c**) 3D topological image of the template PR grating prepared by photolithography. (**d**) Optical microscopic image of the fabricated 2D crystalline grid of $MAPbBr_3$ under illumination by UV light.

Figure 2 shows the microscopic measurements on the practically fabricated structures. Figure 2a–c show the 3D topological mapping of the photoresist grating structures for a period of 50, 20, and 10 μm by a 3D profiler, which verify excellent distribution and homogeneity performance of these grating structures. Although it is not specified in the figures for the clarity of the demonstration, according to the 3D topological measurements, the modulation depth of the gratings or the square cylinder height is about 1 μm for all of the samples. We have also carried out fabrication experiments for a larger grating period of 100 μm and smaller periods than 2 μm to optimize the structural parameters. However, we did not find much difference between the fabrication using 50 and 100 μm. In contrast, for grating periods smaller than 2 μm we did not achieve satisfying crystalline structures of $MAPbBr_3$ on the patterned substrates. Therefore, we include in this work our fabrication and investigation of the patterned $MAPbBr_3$ thin film by three typical periods of 10, 20, and 50 μm.

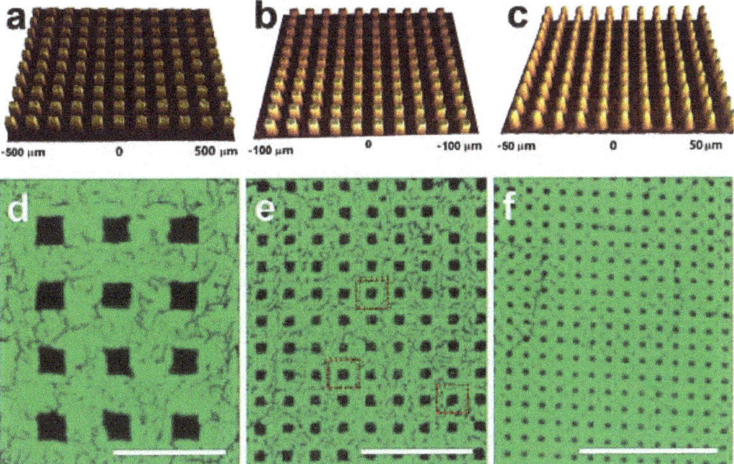

Figure 2. 3D topological images (**a–c**) and optical microscope images (**d–f**) of the template PR grating and the finished crystalline $MAPbBr_3$ grid for periods of 50 μm, 20 μm, and 10 μm, respectively. The scale bars denote a same length scale of 80 μm.

Figure 2d–f show the fluorescence microscope images of the fabricated $MAPbBr_3$ thin film on patterned substrates with photoresist gratings of different periods. For a grating period of 50 μm, due to the large width and large volume of the grating grooves, there are two opposite effects during the crystallization processes. On one hand, a large amount of $MAPbBr_3$ was confined completely into the grating grooves, which favors good crystallization performance inside the large volumes. On the other hand, due to the low concentration of the solution, the materials confined in the grating groove during the spin-coating process are limited; the large volume inside the grating grooves allows more freedom and more time for $MAPbBr_3$ to get crystallized; these factors reduce the quality and homogeneity of the crystals, as can be confirmed by Figure 2d. We can observe that all of the $MAPbBr_3$ are confined to the grooves with very clear edges, however, defects are distributed randomly and extensively in the crystal networks, which can be identified by the dark textures on the bright green background.

Similar effects can still be observed in Figure 2e, where the grating period was reduced to 20 μm. More continuous crystals may be observed in Figure 2e than in Figure 2d, in particular in the space surrounding the photoresist square cylinders, as highlighted by dashed red squares. In contrast, when the grating period is reduced to 10 μm, the continuity and homogeneity of the crystal grid becomes much improved, as can be seen in Figure 2f. The whole network can be observed as a whole crystalline $MAPbBr_3$ film with "black holes" punched by the photoresist square cylinders. The significantly reduced space in the grooves of the grating with a period of 10 μm enables complete filling by $MAPbBr_3$ and high-quality crystallization with continuous connection over the whole grating structure. It is also understandable that breaks may be observed over the large-scale solid structure, as can be verified by the dark defects in Figure 2f. It is thus an indication that for a given solution concentration there exists an optimized channel width for the crystallization of $MAPbBr_3$ in the 2D-distributed grating grooves. For a concentration of 0.14 mg/mL of $MAPbBr_3$ in DMF, the grating period of 10 μm with a rough 50% duty cycle is found to be the best design for producing continuous crystalline structures.

Figure 3a–c show the scanning electron microscope (SEM) images for the samples patterned with a period of 10, 20, and 50 μm, respectively. Figure 3d shows the X-ray diffraction (XRD) spectra for these three samples. According to the SEM images, the sample patterned by the 10-μm grating exhibits the best surface quality among the three samples. Looking at the XRD spectra in Figure 3d at 2θ = 15° in the inset of Figure 3d, we can find that the sample with a 10-μm has the strongest diffraction, the narrowest spectrum, and the smallest diffraction angle. This also implies the best crystalline quality among these three samples. Therefore, the experimental results in Figure 3 again confirmed our analysis above (Figure 2).

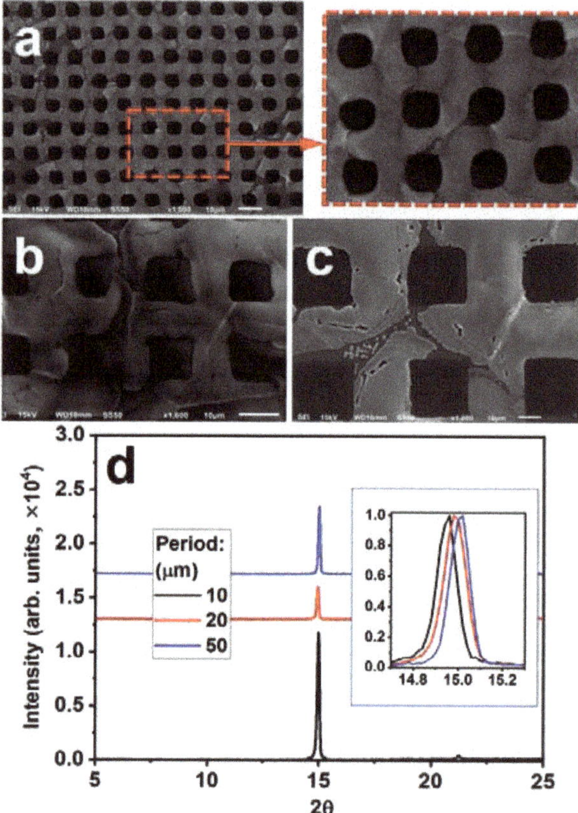

Figure 3. (**a**) SEM image for the patterned structure with a period of 10 μm and an enlarged view of a local area. (**b**,**c**) SEM images for the structures with a patterning period of 20 and 50 μm, respectively. (**d**) XRD spectra measured on the three kinds of structures patterned at different periods. Inset: an enlarged view of the spectra.

3. Optical Spectroscopic Performance of the Gridded Crystalline Film

3.1. Absorption and Photoluminescence

Figure 4a,b show the absorption and emission spectra of the fabricated grating structures, respectively, where the three measured spectra correspond to the samples in Figure 2d–f. According to Figure 4a, the absorption spectra of the patterned structures with different periods are peaked nearly the same at about 525 nm with a bandwidth of about 20 nm at FWHM. However, the three spectra have different contrasts, or they have different background intensity for wavelengths longer than 550 nm. Such background results mainly from the scattering of light by the patterned $MAPbBr_3$. It is understandable by looking at the microscopic images in Figure 2, where the structures have 50 and 20 µm patterning, that they also have more inhomogeneity or defect structures than with 10 µm patterning. Meanwhile, the absorption by the 50 µm grating patterned structures has a broader absorption spectrum due to more defects or inhomogeneity than those with 20- and 10-µm patterning.

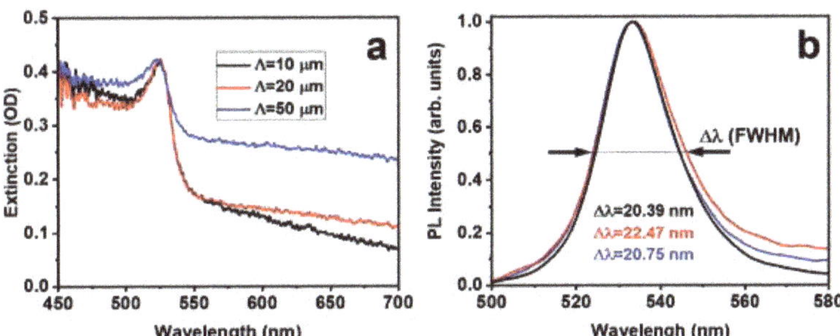

Figure 4. Absorption (**a**) and PL (**b**) spectra of the crystalline $MAPbBr_3$ film gridded at different periods. In (**b**), the bandwidth of each PL spectrum at FWHM is listed.

The PL spectra in Figure 4b are all peaked at about 533 nm, which have nearly the same spectral shape. Nevertheless, they still have slight differences in the bandwidth at FWHM, which is 20.39, 22.47, and 20.75 nm for the grating periods of 10, 20, and 50 µm. Although we did not observe a monotonic variation of the bandwidth when increasing the grating period, we can still identify a narrow emission spectrum at a grating period of 10 µm, implying more homogeneous crystallization of $MAPbBr_3$, which can be further confirmed by the random lasing properties.

3.2. Random Lasing

Figure 5 shows the random lasing performance of the 2D patterned structures at different periods, where Figure 5a,a′, Figure 5b,b′ and Figure 5c,c′ correspond to the grating periods of 50, 20, and 10 µm. In the measurements, a femtosecond laser with a pulse length of 150 fs, a repetition rate of 1 kHz, and a center wavelength of about 400 nm was used as the pump. A spectrometer with a sub-1 nm resolution was positioned in front of the sample and used to measure the emission spectrum. Figure 5a shows the emission spectrum measured at different pump fluence for the sample with a grating period of 50 µm. An obvious random lasing peak occurs as the pump fluence exceeds 48 µJ/cm^2, which increases dramatically with the increasing of the pump fluence. The peak intensity as a function of pump fluence is plotted in Figure 5a′, where two stages can be observed, implying a clear threshold effect of the random lasing performance. Using linear fitting to the measurement data to these two stages, we can resolve a threshold pump fluence lower than 50 µJ/cm^2. Similar measurements and analysis are made for the sample with a patterning period of 20 µm, as shown in Figure 5b,b′. However, a higher pump-fluence threshold can be justified for Λ = 20 µm, which is roughly 60 µJ/cm^2. This can be understood by considering the lower density of defects or higher crystallization quality, as compared with Λ = 50 μm, where the light-scattering strength has been reduced and consequently the optical gain mechanisms were weakened. Similar effects apply further to the case of Λ = 10 μm, as shown in Figure 5c,c′, where random lasing spectrum can be clearly observed when the pump fluence is increased to about 70 μJ/cm^2. This aligns well with the observation of the most homogeneous crystallization structures for Λ = 10 μm. All of the experimental results in Figure 5 have shown clear threshold effects and two-stage variations of the emission intensity with pump fluence, indicating excellent random-lasing performance for all of the samples. In particular, the random lasing performance supplies a further verification of the crystallization qualities at different pattering periods.

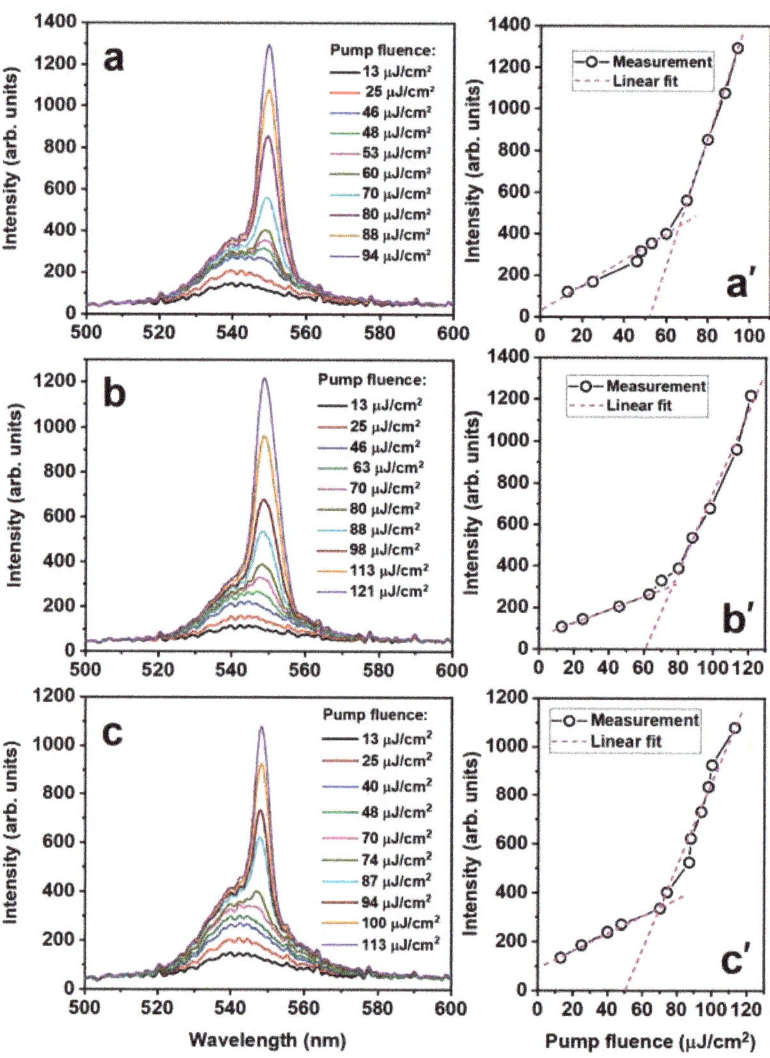

Figure 5. Random lasing performance (emission spectra and lasing threshold) of the $MAPbBr_3$ crystalline gridding at different periods: (**a**) Λ = 50 μm; (**b**) Λ = 20 μm; (**c**) Λ = 10 μm. (**a′–c′**) Emission intensity at about 550 nm as a function of the pump fluence, corresponding to (**a–c**), respectively.

A more convincing analysis is the comparison of the lasing bandwidth. To make this comparison more objective, we need to isolate the random lasing spectrum from the emission spectrum. This can be done using the mathematics below:

$$I_{RL}(\lambda) = I(\lambda) - I_0(\lambda) \times I(535\,\text{nm})/I_0(535\,\text{nm})$$

where $I_{RL}(\lambda)$ is the pure random lasing spectrum, $I(\lambda)$ is the directly measured emission spectrum, $I_0(\lambda)$ is the background fluorescence spectrum at the lowest pump fluence below the threshold for random lasing, and I(535 nm) and I_0(535 nm) are the intensities of the corresponding spectra at 535 nm, which is outside the spectral range of random lasing.

Figure 6 shows the extracted pure random lasing spectra for patterning periods of 50, 20, and 10 μm. There is observed a slight redshift of the peak wavelength (λ_0) of the lasing spectrum with increasing the pattering period. However, we did not observe a monotonic variation of the lasing bandwidth, where we have a full bandwidth at half maximum of 6.42, 7.07, and 4.98 nm for the patterning periods of Λ = 50, 20, and 10 μm, respectively. Nevertheless, for Λ = 10 μm, we achieved the narrowest lasing emission spectrum. This verifies again that the structures with a period of 10 μm have the best crystallization quality and the highest homogeneity, so that the lasing spectrum has a narrowest bandwidth of 4.98 nm.

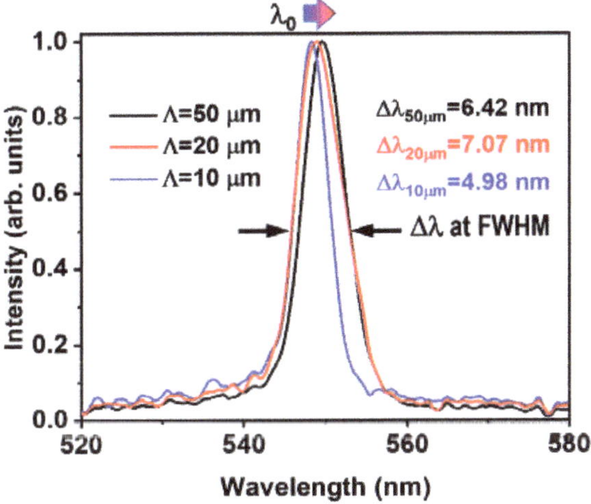

Figure 6. Comparison between the bandwidth of the random lasing spectra measured on the patterned $MAPbBr_3$ crystalline film at different gridding periods.

4. Conclusions

We achieved locally high-quality and large-scale homogeneous crystalline networks of $MAPbBr_3$. This overcomes the challenge of producing large-area thin film devices of hybrid halide perovskites with high quality by spin-coating. At a specific concentration of $MAPbBr_3$ in DMF, the optimized crystal networks were produced in a 2D grating with a period of 10 μm. Random lasing was achieved for all of the two-dimensionally gridded $MAPbBr_3$ crystals with a period of Λ = 10, 20, and 50 μm. However, the lowest lasing threshold was measured for Λ = 50 μm and the narrowest lasing bandwidth was obtained at Λ = 10 μm. The underlying mechanism is that the strong scattering of light by high density of defects with rough interfaces enables the lowest lasing threshold at Λ = 50 μm, while the most homogeneous crystal structures allow random lasing with the narrowest bandwidth at Λ = 10 μm. These discoveries are important for the design of optoelectronic devices based on large-scale $MAPbBr_3$ crystalline films with high quality and high homogeneity.

Author Contributions: Conceptualization, X.Z. and H.X.; methodology, X.Z. and J.H.; validation, X.Z., J.H. and H.X.; formal analysis, X.Z. and J.H.; investigation, J.H. and X.Z.; resources, X.Z.; writing—original draft preparation, X.Z.; writing—review and editing, X.Z.; visualization, X.Z.; supervision, X.Z.; project administration, X.Z.; funding acquisition, X.Z. All authors have read and agreed to the published version of the manuscript.

Funding: This research was funded by National Natural Science Foundation of China, grant number 61735002, 12074020 and by Beijing Municipal Commission of Education, grant number KZ202010005002. The APC was funded by Beijing Municipal Commission of Education.

Conflicts of Interest: The authors declare no conflict of interest.

References

1. Thomas, M.B.; David, A.E.; Leeor, K.; Gary, H.; David, C.H. Hybrid organic–inorganic perovskites: Low-cost semiconductors with intriguing charge-transport properties. *Nat. Rev. Mater.* **2016**, *1*, 15007.
2. Jena, A.K.; Kulkarni, A.; Miyasaka, T. Halide Perovskite Photovoltaics: Background, Status, and Future Prospects. *Chem. Rev.* **2019**, *119*, 3036–3103. [CrossRef]
3. Gu, Z.; Zhou, Z.; Huang, Z.; Wang, K.; Cai, Z.; Hu, X.; Li, L.; Li, M.; Zhao, Y.S.; Song, Y. Controllable Growth of High-Quality Inorganic Perovskite Microplate Arrays for Functional Optoelectronics. *Adv. Mater.* **2020**, *32*, e1908006. [CrossRef]
4. Wang, K.; Xing, G.; Song, Q.; Xiao, S. Micro- and Nanostructured Lead Halide Perovskites: From Materials to Integrations and Devices. *Adv. Mater.* **2020**, *33*, e2000306. [CrossRef]
5. Ahmad, R.; Surendran, A.; Harikesh, P.C.; Haselsberger, R.; Jamaludin, N.F.; John, R.; Koh, T.M.; Bruno, A.; Leong, W.L.; Mathews, N.; et al. Perturbation-Induced Seeding and Crystallization of Hybrid Perovskites over Surface-Modified Substrates for Optoelectronic Devices. *ACS Appl. Mater. Interfaces* **2019**, *11*, 27727–27734. [CrossRef]
6. Cao, L.Z.; Hu, F.R.; Zhang, C.F.; Zhu, S.N.; Xiao, M.; Wang, X.Y. Optical studies of semiconductor perovskite nanocrystals for classical optoelectronic applications and quantum information technoloies: A review. *Adv. Photon.* **2020**, *2*, 054001. [CrossRef]
7. Congreve, D.N.; Weidman, M.C.; Seitz, M.; Paritmongkol, W.; Dahod, N.S.; Tisdale, W.A. Tunable Light-Emitting Diodes Utilizing Quantum-Confined Layered Perovskite Emitters. *ACS Photon.* **2017**, *4*, 476–481. [CrossRef]
8. Cho, H.; Jeong, S.-H.; Park, M.-H.; Kim, Y.-H.; Wolf, C.; Lee, C.-L.; Heo, J.H.; Sadhanala, A.; Myoung, N.; Yoo, S.; et al. Overcoming the electroluminescence efficiency limitations of perovskite light-emitting diodes. *Science* **2015**, *350*, 1222–1225. [CrossRef] [PubMed]
9. Liu, Y.; Cui, J.; Du, K.; Tian, H.; He, Z.; Zhou, Q.; Yang, Z.; Deng, Y.; Chen, D.; Zuo, X.; et al. Efficient blue light-emitting diodes based on quantum-confined bromide perovskite nanostructures. *Nat. Photon.* **2019**, *13*, 760–764. [CrossRef]
10. Yang, W.S.; Park, B.-W.; Jung, E.H.; Jeon, N.J.; Kim, Y.C.; Lee, D.U.; Shin, S.S.; Seo, J.; Kim, E.K.; Noh, J.H.; et al. Iodide management in formamidinium-lead-halide–based perovskite layers for efficient solar cells. *Science* **2017**, *356*, 1376–1379. [CrossRef]
11. Arora, N.; Dar, M.I.; Hinderhofer, A.; Pellet, N.; Schreiber, F.; Zakeeruddin, S.M.; Grätzel, M. Perovskite solar cells with CuSCN hole extraction layers yield stabilized efficiencies greater than 20%. *Science* **2017**, *358*, 768–771. [CrossRef] [PubMed]
12. Ergen, O.; Gilbert, S.M.; Pham, T.; Turner, S.J.; Tan, M.T.Z.; Worsley, M.A.; Zettl, A. Graded bandgap perovskite solar cells. *Nat. Mater.* **2016**, *16*, 522–525. [CrossRef]
13. Kao, T.S.; Chou, Y.-H.; Hong, K.-B.; Huang, J.-F.; Chou, C.-H.; Kuo, H.-C.; Chen, F.-C.; Lu, T.-C. Controllable lasing performance in solution-processed organic–inorganic hybrid perovskites. *Nanoscale* **2016**, *8*, 18483–18488. [CrossRef] [PubMed]
14. Dong, H.; Zhang, C.; Liu, X.; Yao, J.; Zhao, Y.S. Materials chemistry and engineering in metal halide perovskite lasers. *Chem. Soc. Rev.* **2020**, *49*, 951–982. [CrossRef]
15. Gao, Y.; Wang, S.; Huang, C.; Yisheng, G.; Wang, K.; Xiao, S.; Song, Q. Room temperature three-photon pumped $CH_3NH_3PbBr_3$ perovskite microlasers. *Sci. Rep.* **2017**, *7*, 45391. [CrossRef] [PubMed]
16. Duan, Z.; Wang, S.; Yi, N.; Gu, Z.; Gao, Y.; Song, Q.; Xiao, S. Miscellaneous Lasing Actions in Organo-Lead Halide Perovskite Films. *ACS Appl. Mater. Interfaces* **2017**, *9*, 20711–20718. [CrossRef]
17. Gu, Z.; Wang, K.; Sun, W.; Li, J.; Liu, S.; Song, Q.; Xiao, S. Two-Photon Pumped $CH_3NH_3PbBr_3$ Perovskite Microwire Lasers. *Adv. Opt. Mater.* **2015**, *4*, 472–479. [CrossRef]
18. Jeon, T.; Kim, S.J.; Yoon, J.; Byun, J.; Hong, H.R.; Lee, T.-W.; Kim, J.-S.; Shin, B.; Kim, S.O. Hybrid Perovskites: Effective Crystal Growth for Optoelectronic Applications. *Adv. Energy Mater.* **2017**, *7*, 1602596. [CrossRef]
19. Liu, C.; Cheng, Y.-B.; Ge, Z. Understanding of perovskite crystal growth and film formation in scalable deposition processes. *Chem. Soc. Rev.* **2020**, *49*, 1653–1687. [CrossRef]
20. Choi, J.J.; Khan, M.E.; Hawash, Z.; Kim, K.J.; Lee, H.; Ono, L.K.; Qi, Y.B.; Kim, Y.-H.; Park, J.Y. Atomic-scale view of stability and degradation of single-crystal MAPbBr3 surfaces. *J. Mater. Chem. A* **2019**, *7*, 20760. [CrossRef]
21. Wang, K.-H.; Li, L.-C.; Shellaiah, M.; Sun, K.W. Structural and Photophysical Properties of Methylammonium Lead Tribromide ($MAPbBr_3$) Single Crystals. *Sci. Rep.* **2017**, *7*, 1–14. [CrossRef]
22. Zhang, X.P.; Wang, M.; Ma, L.; Fu, Y.L.; Fu, Y.L.; Guo, J.X.; Ma, H.; Zhang, Y.W.; Yan, Z.G.; Han, X.D. Ultrafast two-photon optical switch using singlecrystal hybrid halide perovskites. *Optica* **2021**, *8*, 735–742. [CrossRef]

23. Weng, G.; Tian, J.; Chen, S.; Xue, J.; Yan, J.; Hu, X.; Chen, S.; Zhu, Z.; Chu, J. Giant reduction of the random lasing threshold in $CH_3NH_3PbBr_3$ perovskite thin films by using a patterned sapphire substrate. *Nanoscale* **2019**, *11*, 10636–10645. [CrossRef]
24. Shi, Z.-F.; Sun, X.-G.; Wu, D.; Xu, T.-T.; Tian, Y.-T.; Zhang, Y.-T.; Li, X.-J.; Du, G.-T. Near-infrared random lasing realized in a perovskite $CH_3NH_3PbI_3$ thin film. *J. Mater. Chem. C* **2016**, *4*, 8373–8379. [CrossRef]
25. Wang, Y.-C.; Li, H.; Hong, Y.-H.; Hong, K.-B.; Chen, F.-C.; Hsu, C.-H.; Lee, R.-K.; Conti, C.; Kao, T.S.; Lu, T.-C. Flexible Organometal–Halide Perovskite Lasers for Speckle Reduction in Imaging Projection. *ACS Nano* **2019**, *13*, 5421–5429. [CrossRef] [PubMed]
26. Saouma, F.O.; Stoumpos, C.C.; Wong, J.; Kanatzidis, M.G.; Jang, J.I. Selective enhancement of optical nonlinearity in two-dimensional organic-inorganic lead iodide perovskites. *Nat. Commun.* **2017**, *8*, 742. [CrossRef]
27. Hu, J.; Wang, M.; Tang, F.; Liu, M.; Mu, Y.; Fu, Y.; Guo, J.; Song, X.; Zhang, X. Threshold Size Effects in the Patterned Crystallization of Hybrid Halide Perovskites for Random Lasing. *Adv. Photon. Res.* **2020**, *2*, 2000097. [CrossRef]
28. Oded, M.; Kelly, S.T.; Gilles, M.K.; Mueller, A.H.E.; Shenhar, R. Periodic nanoscale patterning of polyelectrolytes over square centimeter areas using block copolymer templates. *Soft Matter* **2016**, *12*, 4595–4602. [CrossRef] [PubMed]
29. Coutts, M.J.; Zareie, H.M.; Cortie, M.; Phillips, M.; Wuhrer, R.; McDonagh, A. Exploiting Zinc Oxide Re-emission to Fabricate Periodic Arrays. *ACS Appl. Mater. Interfaces* **2010**, *2*, 1774–1779. [CrossRef] [PubMed]
30. Petit, C.A.P.; Carbeck, J.D. Combing of nolecules in microchannels (COMMIC): A method for micropatterning and orienting stretched molecules of DNA on a surface. *Nano Lett.* **2003**, *3*, 1141. [CrossRef]
31. Liu, J.; Zhang, X.; Li, W.; Jiang, C.; Wang, Z.; Xiao, X. Recent progress in periodic patterning fabricated by self-assembly of colloidal spheres for optical applications. *Sci. China Mater.* **2020**, *63*, 1418–1437. [CrossRef]

Article

Photodetector Based on $CsPbBr_3/Cs_4PbBr_6$ Composite Nanocrystals with High Detectivity

Yue Han, Rong Wen *, Feng Zhang, Linlin Shi, Wenyan Wang, Ting Ji, Guohui Li, Yuying Hao, Lin Feng * and Yanxia Cui

College of Physics and Optoelectronics, Key Laboratory of Advanced Transducers and Intelligent Control System of Ministry of Education, Key Laboratory of Interface Science and Engineering in Advanced Materials of Ministry of Education, Taiyuan University of Technology, Taiyuan 030024, China; yuehan2020310158@163.com (Y.H.); zhangfeng@tyut.edu.cn (F.Z.); shilinlinsll@sina.cn (L.S.); wangwenyan@tyut.edu.cn (W.W.); jiting@tyut.edu.cn (T.J.); liguohui@tyut.edu.cn (G.L.); haoyuying@tyut.edu.cn (Y.H.); yanxiacui@gmail.com (Y.C.)
* Correspondence: wenrong@tyut.edu.cn (R.W.); fenglin@tyut.edu.cn (L.F.)

Abstract: High-quality, all-inorganic $CsPbBr_3/Cs_4PbBr_6$ composite perovskite nanocrystals (NCs) were obtained with all-solution-processing at room temperature, and a photodetector (PD) with high detectivity was realized based on $CsPbBr_3/Cs_4PbBr_6$ NCs. The detectivity (D^*) of the proposed PD is 4.24×10^{12} Jones under 532 nm illumination, which is among the highest levels for PDs based on all-inorganic perovskite NCs. In addition, a high linear dynamic range (LDR) of 115 dB under 1 V bias was also realized. Furthermore, the underlying mechanism for the enhanced performance of the proposed PD was discussed. Our work might promote the preparation of high-performance PDs based on dual-phase all-inorganic perovskite nanocrystals.

Keywords: all-inorganic lead halide perovskite; $CsPbBr_3$; Cs_4PbBr_6; perovskite nanocrystals; photodetector

Citation: Han, Y.; Wen, R.; Zhang, F.; Shi, L.; Wang, W.; Ji, T.; Li, G.; Hao, Y.; Feng, L.; Cui, Y. Photodetector Based on $CsPbBr_3/Cs_4PbBr_6$ Composite Nanocrystals with High Detectivity. *Crystals* **2021**, *11*, 1287. https://doi.org/10.3390/cryst11111287

Academic Editor: Luisa De Marco

Received: 4 October 2021
Accepted: 21 October 2021
Published: 24 October 2021

1. Introduction

All-inorganic lead halide perovskites have shown excellent photoelectrical properties with high absorption coefficients, wide spectra, high carrier mobility values, and long charge diffusion lengths. Because of these advantages, they are widely used in PDs [1–4], solar cells [5–8], lasers [9–11], light emitting diodes [12–14], and so on. In addition, compared with the organic–inorganic hybrid perovskite, all-inorganic lead halide perovskites have higher chemical stability to the environment because there are no organic ions in them. Among many kinds of all-inorganic lead halide perovskites, $CsPbBr_3$ NCs demonstrate strong competitiveness for their excellent photoelectric characteristics, such as adjustable photoluminescence (PL) with narrow full-width at half maximum (FWHM) in visible light, an easy preparation process for their devices, and small size which causes a quantum size effect. Moreover, the $CsPbBr_3$ NCs can be stored in air for 30 days with no obvious decrease in their photoelectric performance [15]. Some research on optoelectronic devices based on $CsPbBr_3$ NCs has been conducted. For example, Li et al. reported an interesting recyclable dissolution–recrystallization phenomenon of $CsPbBr_3$ NCs and its application on the room temperature self-healing of compact and smooth carrier channels for PDs with D^* of 6.1×10^{10} Jones [16]. Zeng et al. demonstrated a high-performance self-actuation PD, for which $CsPbBr_3$ NCs were synthesized using room temperature saturation recrystallization, and the NC films were prepared by a centrifugal casting method. Compared with the drip-coating method, the photocurrent of the device increased three-fold, and the optimized device had a high on/off ratio (>10^5) [15]. However, the D^* of these all-inorganic perovskite NC PDs is not high enough. Generally, the detectivity can be improved by increasing the bright current and suppressing the dark current. Additionally, improving the quality of nanocrystalline films to reduce defect states is vital for the suppression of the dark current.

Quan et al. have found that embedding $CsPbBr_3$ NCs in the Cs_4PbBr_6 matrix can prevent the agglomeration of $CsPbBr_3$ NCs [17]. This means that the presence of Cs_4PbBr_6 NCs could improve the quality of NC films. Thus, we prepared $CsPbBr_3/Cs_4PbBr_6$ composite NCs by controlling the proportion of $PbBr_2$ in the synthesis process. Cs_4PbBr_6 is an indirect band gap semiconductor with a wide band gap (3.8 eV) [18], and Cs_4PbBr_6 NC is a zero-dimensional material, which absorbs light yet does not emit it [19]. This non-luminous property can effectively reduce radiation recombination. In addition, it has been proved that the presence of Cs_4PbBr_6 will enhance the absorption of the dual-phase perovskite NCs in the UV spectrum [17,20,21]. This is helpful for improving the detection performance of the PD in the UV spectrum.

Taking the above-mentioned issues into account, the saturation recrystallization method was utilized to synthesize the dual-phase NCs at room temperature. The dual-phase NC film was spin-coated on the customized substrate and Au electrode. Moreover, the planar metal-semiconductor-metal (MSM) PDs based on $CsPbBr_3/Cs_4PbBr_6$ composite NCs were realized. The proposed PD exhibited a high D^* of 4.24×10^{12} Jones, with an LDR of 115 dB under 1 V bias. To explain the improvement of D^*, the role of Cs_4PbBr_6 NCs was investigated. Our work might provide a guidance for developing a high-performance PD based on $CsPbBr_3/Cs_4PbBr_6$ perovskite NCs.

2. Materials and Methods

Materials: Cesium bromide (CsBr, 99.5%), lead bromide ($PbBr_2$, 99.0%), Oleic acid (OA, 85%) and oleylamine (OAm, 80–90%) were purchased from Aladdin. Toluene (AR) was purchased from Beijing Chemical Works. All these chemicals were used as received.

Synthesis of $CsPbBr_3/Cs_4PbBr_6$ NCs: The saturation recrystallization method was utilized to synthesize the NCs at room temperature without any noble gases, and the specific synthesis steps are shown in Figure 1. Firstly, the beaker and the small glass bottle were each cleaned with deionized water, isopropanol, and absolute ethanol for 15 min, and then dried with high-purity nitrogen. After that, they were placed on a hot table at 80 °C to accelerate solvent evaporation and ensure no residual water. Then, 0.4 mmol of CsBr and 0.4 mmol of $PbBr_2$ were dissolved in 10 mL of DMSO. A clear and transparent solution was obtained by ultrasonic cleaning, and no water entered the solution in this process. Then, 0.5 mL of OAm and 1 mL of OA were added as ligands of NCs to prepare a precursor solution. After that, 1 mL of the above precursor solution was dribbled into the 10 mL of toluene solution with a fast speed of 800 r/min. In this way, various inorganic ions were transferred from the benign solvent to the undesirable solvent, and the NCs were deposited due to the much higher solubility of NCs in DMSO than in DMSO mixed with toluene.

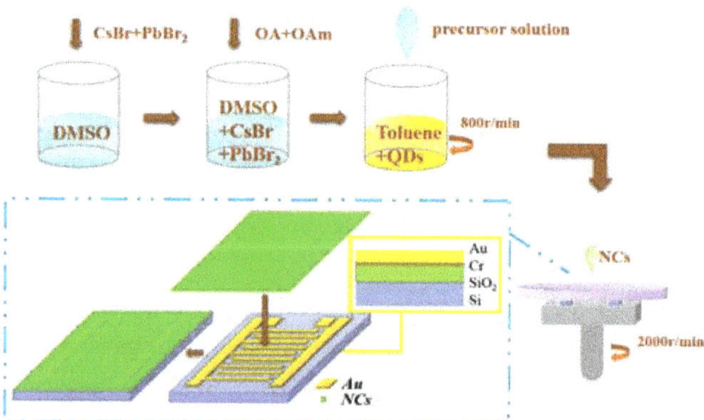

Figure 1. Schematic diagram of the fabrication process of the PD based on $CsPbBr_3/Cs_4PbBr_6$ NCs.

Fabrication of PD: The fabrication of the device was mainly based on the gold interdigital electrode and the inorganic perovskite NCs. The schematic diagram of the fabrication process is illustrated in Figure 1. The interdigital space of the gold interdigital electrode is 3 μm and the thickness of the Au electrode is about 100 $\pm$ 10 nm. The substrate is made up of silicon (Si) wafer, SiO_2, and chromium (Cr) plating. A 300 nm-thick SiO_2 layer covers the surface of the silicon wafer. Additionally, there is a Cr plating layer with a thickness of 10 nm between the SiO_2 layer and the Au electrode. The planar PD was fabricated in a glove box by spin-coating. The NC solution was spin-coated on the substrate with the rotational speed of 2000 r/min. The device was then placed on a hot platform at 90 °C to vaporize the methylbenzene and DMSO.

Characteristics of nanocrystals: The photoelectric property of NCs was tested by a light-emitting diode (LED) light source (375 nm, Thorlabs, Shanghai, China). The optical image of NCs was characterized using an optical microscope (Nikon, LV150, Shanghai, China). The high-resolution transmission electron microscopy (HRTEM) images of NCs were collected using a JEOL JEM-2100 microscope (Nippon Electronics Corporation, Shanghai, China) at an accelerating voltage of 100 kV. The X-ray diffraction (XRD) spectrum of the NCs was measured by a diffractometer (Haoyuan Instrument Co., Ltd., DX-2700, Dandong, China). The absorption spectrum of NCs was determined by an ultraviolet-visible absorption spectrometer (Shimadzu, UV-2600, Shimadzu, Japan). The photoluminescence spectrum at room temperature was carried out by a home-built fluorescence spectrophotometer system using a 343 nm femtosecond laser (Light Conversion, Carbide 5W, Vilnius, Lithuania) as the excitation light source.

Photoelectric characteristics: The PD was placed on the probe table (Prcbe, Mini), and the external shielding box was used to isolate the external optical signals throughout the test. The current–voltage (*I–V*) curves and the transient photo responses of the PD based on inorganic perovskite NCs were tested using a semiconductor analyzer (Agilent, B1500, Shanghai, China). For completing the *I–V* measurements, the light intensity was tuned by adjusting the LED light source (375 nm, Thorlabs, Shanghai, China) at a fixed driving voltage. The transient photo responses were acquired with an LED lamp, which emitted light faster than the response of the measured PDs.

3. Results and Discussion

3.1. Composition of NCs

To determine the composition of NCs, XRD measurements were carried out, as shown in Figure 2a. The peaks at 15.185°, 21.551°, 30.644°, 34.371° and 37.767° can be indexed to the (100), (110), (200), (210) and (211) lattice planes, respectively, which are the characteristic peaks of the $CsPbBr_3$ phase. On the other hand, the peaks at 12.885°, 20.079°, 22.413°, 25.427°, 27.510°, 28.603° and 30.268° can be identified as the (110), (113), (300), (024), (131), (214) and (223) reflections, respectively, which are the characteristic peaks of Cs_4PbBr_6 [18,21]. In addition, the intensity of the diffraction peaks of Cs_4PbBr_6 is much higher than those of $CsPbBr_3$, which means that the main component of the NCs is Cs_4PbBr_6. To further investigate the composition of the NCs, the HRTEM images were obtained and shown in Figure 2b, and the selected area corresponding to Figure 2b is illustrated in Figure 2c. It can be found that the lattice fringes of 0.68 nm can be indexed to the (110) lattice planes of orthorhombic Cs_4PbBr_6, while the lattice fringes of 0.29 nm can be indexed to the (200) lattice planes of cubic $CsPbBr_3$. Thus, the dual-phase was also evidenced by the selected area TEM of dual-phase NCs.

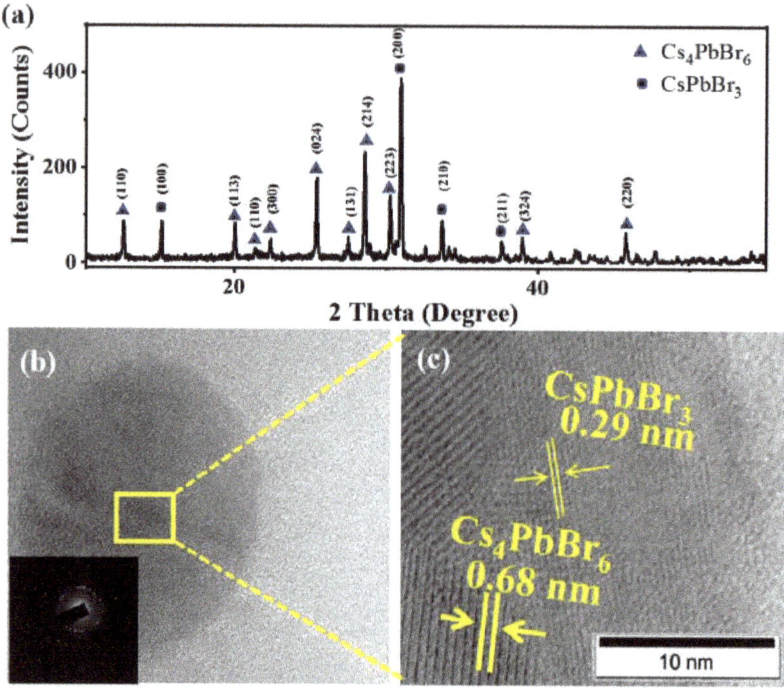

Figure 2. (**a**) XRD patterns of the dual-phase $CsPbBr_3/Cs_4PbBr_6$ NCs; (**b**) HRTEM image of the dual-phase inorganic perovskite NC film, inset: Fast Fourier transform (FFT) image of the dual-phase inorganic perovskite NC film; (c) HRTEM corresponding to Figure 2b.

3.2. Optical Property of NCs

In order to investigate the optical property of NCs, its absorption and PL spectrum were tested, as shown in Figure 3a,b. We found that there were two absorption peaks in the absorption spectrum. One peak was located at 505 nm, and the other one was located at 316 nm. According to the existing research on $CsPbBr_3$, we know that the absorption peak at 505 nm is attributed to $CsPbBr_3$. Thus, it can be inferred that the absorption peak at 316 nm was caused by Cs_4PbBr_6. In addition, the absorption band edges of $CsPbBr_3$ and Cs_4PbBr_6 NCs are 360 nm and 547 nm, respectively. Based on this, the band gaps of $CsPbBr_3$ and Cs_4PbBr_6 NCs were calculated, which were 2.3 eV and 3.8 eV, respectively. This is in agreement with the literature [18]. For the PL spectrum (shown in Figure 3b), a narrow emission peak was observed at 520 nm with a FWHM of 26 nm, which is 3.1 nm lower that of pure $CsPbBr_3$ NCs (29.1 nm) [22].This suggests that the quality of NC films was improved by the introduction of Cs_4PbBr_6 NCs. It is worth noting that there was only one PL peak that originated from $CsPbBr_3$ NCs, which is consistent with the previous studies' finding that Cs_4PbBr_6 does not emit light [23]. Moreover, the PL peak was in the green light emission band, which echoes the luminescence of the sample shown in the inset of Figure 3b. $CsPbBr_3/Cs_4PbBr_6$ NCs in toluene (illustration) were yellow under sunlight. However, when exposed to ultraviolet light, the $CsPbBr_3/Cs_4PbBr_6$ NCs in toluene emitted bright green light. This is also consistent with the reported emission band of $CsPbBr_3/Cs_4PbBr_6$ NCs [21].

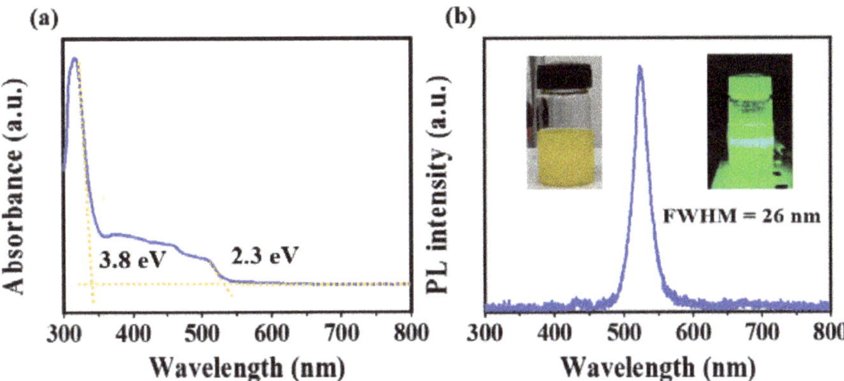

Figure 3. (**a**) Absorption spectrum of $CsPbBr_3/Cs_4PbBr_6$ NCs; (**b**) PL spectrum of $CsPbBr_3/Cs_4PbBr_6$ NCs, insets show NC solution illuminated by a fluorescent lamp and a 375 nm LED lamp, respectively.

3.3. Performance of the Composite NC PD

The dual-phase inorganic perovskite NC film was spin-coated on the gold interdigital electrodes, and the planar MSM PDs were prepared. The *I–V* curves of the proposed PD under 375 nm LED illumination with different light intensities are shown in Figure 4a. It can be found that the photocurrent increased with the rise in the optical power density. Moreover, the dark current at 1 V bias was 6.67 pA, which was significantly suppressed compared with that of the PD based on pure $CsPbBr_3$ NCs (1.5 nA at 2 V bias) [15]. It suggests that the quality of NC film was improved by the introduction of Cs_4PbBr_6 NCs, which could prevent the agglomeration of $CsPbBr_3$ NCs. The light current with a light intensity of 10.2 mW/cm^2 at 1V bias was 19.32 nA and the ratio of light current to dark current was 2894 at 1 V bias. This value was much lower than that of the PD based on $CsPbBr_3$ NCs (10^5 under 4.65 mW/cm^2 at 2 V bias), which meant that the bright current was weakened by the introduction of Cs_4PbBr_6 NCs. The external quantum efficiency (η_{EQE}), responsivity (*R*), and *D** of the proposed PD were calculated and the results are shown in Figure 4c–e. As the power density of the incident light increased from 13 μW/cm^2 to 2689 mW/cm^2, the η_{EQE} decreased continuously. At the lowest detectable illumination of 13 μW/cm^2 at a bias of 1V, the η_{EQE} was only 22.1%. This value is also lower than the PDs based on $CsPbBr_3$ NCs (41%) [16]. These phenomena may be due to the fact that the main component of dual-phase NC film is Cs_4PbBr_6. Moreover, the wide band gap of Cs_4PbBr_6 determines that the absorption wavelength of Cs_4PbBr_6 must be lower than 360 nm. However, due to the limitation of the experimental conditions, our test wavelength could only reach 375 nm. Thus, the light absorption intensity and the amount of photogenerated carriers of the PD based on $CsPbBr_3/Cs_4PbBr_6$ composite NCs were correspondingly reduced under the illumination of 375 nm. This also led to low *R*, which was only 0.094 A/W under the optical power of 13 μW/cm^2. The expression of *R* was $\frac{I_L-I_d}{P_{in}S}$ (I_L is photocurrent, I_d is dark current, P_{in} is incident optical power density, *S* is effective area). Although the PD had an ultra-low dark current, the light current was not satisfied, which led to a small *R*. However, compared with the previous reported MSM PDs based on $CsPbBr_3$ NCs (summarized in Table 1), the *D** of the proposed PD was among the highest level, which reached 4.24 × 10^{12} Jones under the optical power of 13 μW/cm^2. Due to the great improvement of D*, it could be expected that the LDR would also be improved. Here, the LDR of the proposed PD under 1 V bias with 532 nm continuous laser illumination were shown in Figure 4b. It can be found that the proposed PD was capable of detecting incident light as low as 13 μW/cm^2 under 1 V bias, and its LDR was 115 dB.

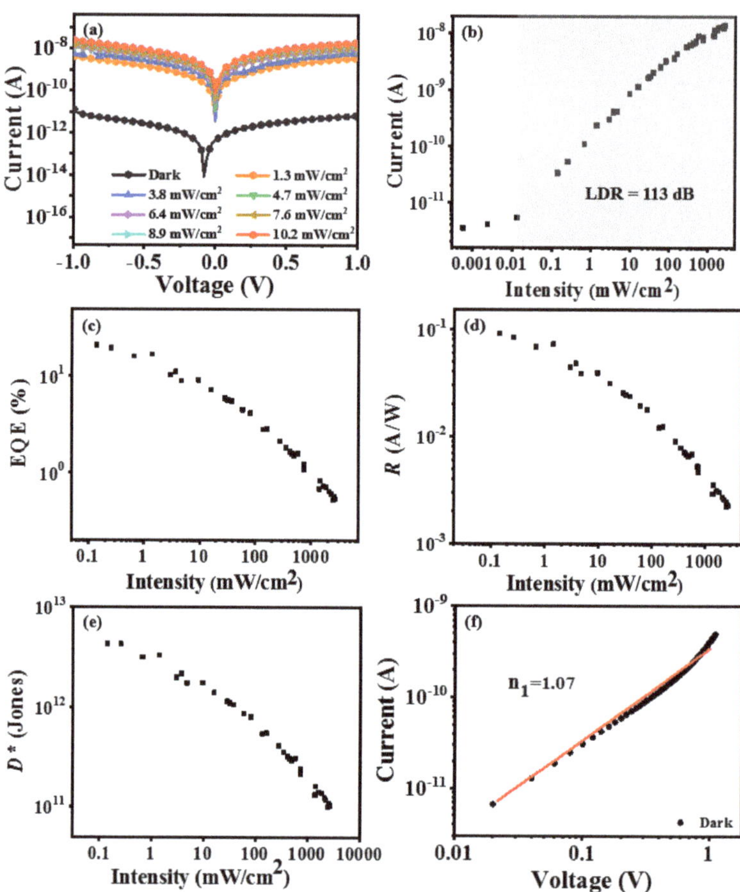

Figure 4. (**a**) *I*–*V* curves of $CsPbBr_3/Cs_4PbBr_6$ NC PD with different optical power densities under the irradiation of a 375 nm LED lamp; (**b**–**e**) LDR, η_{EQE}, *R*, and *D** measured using a 532 nm laser as a light source; (**f**) Change trend of the dark current after taking the logarithm of the x-axis and y-axis, the red line is the fitting curve.

Table 1. Summary of the performance of inorganic perovskite NC PDs.

Nanocrystal	*R* (mA/W)	η_{EQE} (%)	*D** (Jones)	T_{rise}/T_{fall} (ms)	Ref.
$CsPbI_3$	-	-	-	24 / 29	[23]
$CsPbBr_3$	0.18	41	6.1×10^{10}	1.8 / 1.0	[16]
$CsPbCl_3$	1890	-	-	41 / 43	[24]
$CsPbBr_3$	4.71	16.69	4.56×10^{8}	0.2 / 1.3	[5]
$CsPbBr_3/Cs_4PbBr_6$	0.094	22.1	4.24×10^{12}	10.85 / 2.85	Our work

In order to explain the reason for the improvement of *D**, the type of contact between the semiconductor and the metal electrode was analyzed. Due to the high work function of Au, it can be expected that the interface between the inorganic perovskite NCs and the Au electrode will form a good ohmic contact [15]. This prediction was proved by the experiment. The *I*–*V* curve at different biases were measured, and Figure 4f shows the change trend of the dark current after taking the logarithm of the x-axis and y-axis. It is clear that the current increased significantly with the increase in bias, and the slope (n_1) of the curve is 1.07, which is almost linear. This means that a good ohmic contact was formed between the dual-phase NCs and Au electrode, providing a guarantee for the rapid transfer of the carrier. On the other hand, the band gaps of $CsPbBr_3$ and Cs_4PbBr_6 were 2.3 eV and 3.8 eV, respectively, and both the VBM and CBM of Cs_4PbBr_6 were higher than those of $CsPbBr_3$. Thus, by combining $CsPbBr_3$ NCs with Cs_4PbBr_6 NCs, the energy band of the composite NCs would be moved up as a whole relative to the Au electrode. Moreover, the height difference between the VBM of the composite NCs and the Fermi energy of the Au electrode would be reduced, which facilitates the transmission of electrons. This is also helpful for the improvement of D*. In addition, both the strongly suppressed dark current and the narrowed FWHM of the PL spectrum proved that the quality of the NC film was indeed greatly improved with the introduction of Cs_4PbBr_6 NCs. Furthermore, the radiation recombination could also be suppressed due to the non-luminous property of Cs_4PbBr_6 NCs. All these behaviors led to the improvement of D^*.

To study the photoresponse of the composite NC PD, the transient response performance under 375 nm with a light intensity of 10.2 mW/cm^2 at 1 V bias were measured (shown in Figure 5a), which indicated that the proposed PD could respond stably when the illumination was turned on and off. Moreover, response speed was another important index of the photodetector, and the transient photocurrent response relationship (*I-T*) curve of the proposed PD under 1 V bias is shown in Figure 5b. According to the definitions of the rise time (T_{rise}) and fall time (T_{fall}) for the time when the photocurrent increased from 10% to 90% (declined from 90% to 10%) in the on and off cycles under light, T_{rise} and T_{fall} were 10.85 ms and 2.25 ms, respectively. The sum of the rise time and the fall time was considered the response time of the PD, which was 13.1 ms.

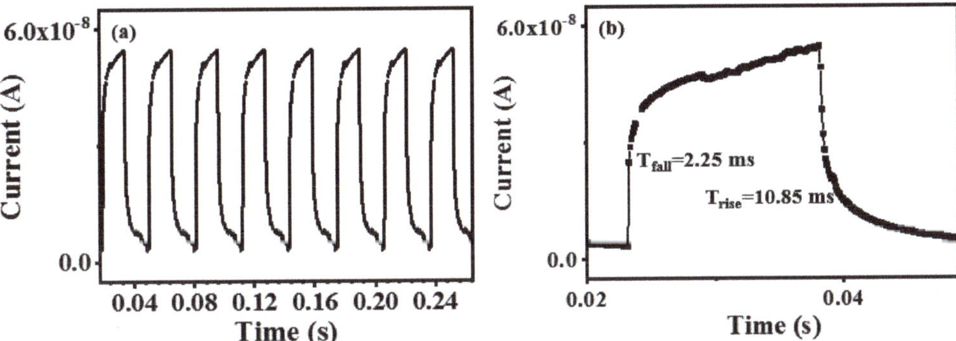

Figure 5. (**a**) Photocurrent–time response measured under 375 nm with a light intensity of 10.2 mW/cm^2; (**b**) Rise and fall time of the proposed PD.

4. Conclusions

In summary, we have successfully demonstrated all-solution-processed PDs based on $CsPbBr_3/Cs_4PbBr_6$ composite NCs. A good ohmic contact was formed between the gold electrode and $CsPbBr_3/Cs_4PbBr_6$ composite NCs, which provided a guarantee for the rapid transfer of the carrier. In addition, both the strongly suppressed dark current and the narrowed FWHM of the PL spectrum proved that the quality of the NC film was greatly improved with the introduction of Cs_4PbBr_6 NCs. Moreover, the non-luminous property of Cs_4PbBr_6 NCs could inhibit the radiation recombination. Thus, the D^* of the PD based on $CsPbBr_3/Cs_4PbBr_6$ composite NCs was improved, reaching 4.24×10^{12} Jones. Our work might provide guidance for developing a high-performance PD based on $CsPbBr_3/Cs_4PbBr_6$ perovskite NCs.

Author Contributions: Conceptualization, R.W. and L.F.; formal analysis, G.L. and Y.H. (Yuying Hao); funding acquisition, W.W., T.J., G.L., Y.H. (Yuying Hao) and Y.C.; investigation, Y.H. (Yue Han) and Y.C.; methodology, F.Z. and T.J.; project administration, R.W. and L.F.; resources, Y.C.; supervision, Y.C.; validation, Y.H. (Yue Han) and L.S.; writing–original draft, Y.H. (Yue Han); writing–review and editing, R.W. and L.F. All authors have read and agreed to the published version of the manuscript.

Funding: This work was funded by the National Natural Science Foundation of China (No. 61922060, 61775156, 61805172, 61905173, 62174117, 12104334), the Key Research and Development (International Cooperation) Program of Shanxi Province (201803D421044), the Henry Fok Education Foundation Young Teachers Fund, the Transformation Cultivation Project of the University of Scientific and Technological Achievements of Shanxi Province (2020CG013), the Research Project Supported by the Shanxi Scholarship Council of China (2021-033) and the Special Project of Introduced Talents of Lvliang City.

Conflicts of Interest: The authors declare no conflict of interest.

References

1. Wu, L.Z.; Hu, H.C.; Xu, Y.; Jiang, S.; Chen, M.; Zhong, Q.X.; Yang, D.; Liu, Q.P.; Zhao, Y.; Sun, B.Q.; et al. From Nonluminescent Cs_4PbX_6 (X = Cl, Br, I) Nanocrystals to Highly Luminescent $CsPbX_3$ Nanocrystals: Water-Triggered Transformation through a CsX-Stripping Mechanism. *Nano Lett.* **2017**, *17*, 5799–5804. [CrossRef] [PubMed]
2. Pan, R.; Li, H.Y.; Wang, J.; Jin, X.; Li, Q.H.; Wu, Z.M.; Gou, J.; Jiang, Y.D.; Song, Y.L. High-Responsivity Photodetectors Based on Formamidinium Lead Halide Perovskite Quantum Dot-Graphene Hybrid. *Part. Part. Syst. Char.* **2018**, *35*, 9. [CrossRef]
3. Xue, J.; Zhu, Z.F.; Xu, X.B.; Gu, Y.; Wang, S.L.; Xu, L.M.; Zou, Y.S.; Song, J.Z.; Zeng, H.B.; Chen, Q. Narrowband Perovskite Photodetector-Based Image Array for Potential Application in Artificial Vision. *Nano Lett.* **2018**, *18*, 7628–7634. [CrossRef] [PubMed]
4. Wang, W.Y.; Shi, L.L.; Zhang, Y.; Li, G.H.; Hao, Y.Y.; Zhu, F.R.; Wang, K.Y.; Cui, Y.X. Effect of photogenerated carrier distribution on performance enhancement of photomultiplication organic photodetectors. *Org. Electron.* **2019**, *68*, 56–62. [CrossRef]
5. Liang, J.; Wang, C.X.; Wang, Y.R.; Xu, Z.R.; Lu, Z.P.; Ma, Y.; Zhu, H.F.; Hu, Y.; Xiao, C.G.; Yi, X.; et al. All-Inorganic Perovskite Solar Cells. *J. Am. Chem. Soc.* **2016**, *138*, 15829–15832. [CrossRef]
6. Kulbak, M.; Gupta, S.; Kedem, N.; Levine, I.; Bendikov, T.; Hodes, G.; Cahen, D. Cesium Enhances Long-Term Stability of Lead Bromide Perovskite-Based Solar Cells. *J. Phys.Chem. Lett.* **2016**, *7*, 167–172. [CrossRef]
7. Eperon, G.E.; Paternò, G.M.; Sutton, R.J.; Zampetti, A.; Haghighirad, A.A.; Cacialli, F.; Snaith, H.J. Inorganic caesium lead iodide perovskite solar cells. *J. Mater. Chem. A* **2015**, *3*, 19688–19695. [CrossRef]
8. Shi, L.L.; Cui, Y.X.; Gao, Y.P.; Wang, W.Y.; Zhang, Y.; Zhu, F.R.; Hao, Y.Y. High Performance Ultrathin MoO_3/Ag Transparent Electrode and Its Application in Semitransparent Organic Solar Cells. *Nanomaterials* **2018**, *8*, 473. [CrossRef] [PubMed]
9. Li, G.H.; Che, T.; Ji, X.Q.; Liu, S.D.; Hao, Y.Y.; Cui, Y.X.; Liu, S.Z. Record-Low-Threshold Lasers Based on Atomically Smooth Triangular Nanoplatelet Perovskite. *Adv. Funct. Mater.* **2019**, *29*, 1805553. [CrossRef]
10. Wang, Y.; Li, X.M.; Song, J.Z.; Xiao, L.; Zeng, H.B.; Sun, H.D. All-Inorganic Colloidal Perovskite Quantum Dots: A New Class of Lasing Materials with Favorable Characteristics. *Adv. Mater.* **2015**, *27*, 7101–7108. [CrossRef]
11. Wang, Y.; Li, X.M.; Zhao, X.; Xiao, L.; Zeng, H.B.; Sun, H.D. Nonlinear Absorption and Low-Threshold Multiphoton Pumped Stimulated Emission from All-Inorganic Perovskite Nanocrystals. *Nano Lett.* **2016**, *16*, 448–453. [CrossRef] [PubMed]
12. Gao, C.H.; Yu, F.X.; Xiong, Z.Y.; Dong, Y.J.; Ma, X.J.; Zhang, Y.; Jia, Y.L.; Wang, R.; Chen, P.; Zhou, D.Y.; et al. 47-Fold EQE improvement in $CsPbBr_3$ perovskite light-emitting diodes via double-additives assistance. *Org. Electron.* **2019**, *70*, 264–271. [CrossRef]
13. Shi, Z.F.; Li, Y.; Zhang, Y.T.; Chen, Y.S.; Li, X.J.; Wu, D.; Xu, T.T.; Shan, C.X.; Du, G.T. High-Efficiency and Air-Stable Perovskite Quantum Dots Light-Emitting Diodes with an All-Inorganic Heterostructure. *Nano Lett.* **2017**, *17*, 313–321. [CrossRef] [PubMed]
14. Cho, H.; Jeong, S.H.; Park, M.H.; Kim, Y.H.; Wolf, C.; Lee, C.L.; Heo, J.H.; Sadhanala, A.; Myoung, N.; Yoo, S.; et al. Overcoming the electroluminescence efficiency limitations of perovskite light-emitting diodes. *Science* **2017**, *350*, 1222–1225. [CrossRef] [PubMed]
15. Dong, Y.H.; Gu, Y.; Zou, Y.S.; Song, J.Z.; Xu, L.M.; Li, J.H.; Xue, J.; Li, X.M.; Zeng, H.B. Improving All-Inorganic Perovskite Photodetectors by Preferred Orientation and Plasmonic Effect. *Small* **2016**, *12*, 5622–5632. [CrossRef] [PubMed]
16. Li, X.; Yu, D.; Cao, F.; Gu, Y.; Wei, Y.; Wu, Y.; Song, J.Z.; Zeng, H.B. Healing All-Inorganic Perovskite Films via Recyclable Dissolution-Recyrstallization for Compact and Smooth Carrier Channels of Optoelectronic Devices with High Stability. *Adv. Funct. Mater.* **2016**, *26*, 5903–5912. [CrossRef]
17. Quan, L.N.; Quintero-Bermudez, R.; Voznyy, O.; Walters, G.; Jain, A.; Fan, J.Z.; Zheng, X.; Yang, Z.; Sargent, E.H. Highly Emissive Green Perovskite Nanocrystals in a Solid State Crystalline Matrix. *Adv. Mater.* **2017**, *29*, 1605945. [CrossRef]
18. Tong, G.Q.; Li, H.; Zhu, Z.F.; Zhang, Y.; Yu, L.W.; Xu, J.; Jiang, Y. Enhancing Hybrid Perovskite Detectability in the Deep Ultraviolet Region with Down-Conversion Dual-Phase ($CsPbBr_3$-Cs_4PbBr_6) Films. *J. Phys. Chem. Lett.* **2018**, *9*, 1592–1599. [CrossRef] [PubMed]

19. Wang, L.L.; Liu, H.; Zhang, Y.H.; Mohammed, O.F. Photoluminescence Origin of Zero-Dimensional Cs_4PbBr_6 Perovskite. *ACS Energy Lett.* **2019**, *5*, 87–99. [CrossRef]
20. Ling, Y.C.; Tan, L.; Wang, X.; Zhou, Y.; Xin, Y.; Ma, B.; Hanson, K.; Gao, H.W. Composite Perovskites of Cesium Lead Bromide for Optimized Photoluminescence. *J. Phys .Chem. Lett.* **2017**, *8*, 3266–3271. [CrossRef]
21. Jing, Q.; Xu, Y.; Su, Y.C.; Xing, X.; Lu, Z.D. A systematic study of the synthesis of cesium lead halide nanocrystals: Does Cs_4PbBr_6 or $CsPbBr_3$ form? *Nanoscale* **2019**, *11*, 1784–1789. [CrossRef] [PubMed]
22. Shen, K.; Li, X.; Xu, H.; Wang, M.Q.; Dai, X.; Guo, J.; Zhang, T.; Li, S.B.; Zou, G.F.; Choy, K.L.; et al. Enhanced performance of ZnO nanoparticle decorated all-inorganic $CsPbBr_3$ quantum dot photodetectors. *J. Mater. Chem. A* **2019**, *7*, 6134–6142. [CrossRef]
23. Ramasamy, P.; Lim, D.H.; Kim, B.; Lee, S.H.; Lee, M.S.; Lee, J.S. All-inorganic cesium lead halide perovskite nanocrystals for photodetector applications. *Chem. Commun.* **2016**, *52*, 2067–2070. [CrossRef] [PubMed]
24. Zhang, J.G.; Wang, Q.; Zhang, X.S.; Jiang, J.X.; Gao, Z.F.; Jin, Z.W.; Liu, S.Z. High-performance transparent ultraviolet photodetectors based on inorganic perovskite $CsPbCl_3$ nanocrystals. *RSC Adv.* **2017**, *7*, 36722–36727. [CrossRef]

MDPI

Review

Review on Y6-Based Semiconductor Materials and Their Future Development via Machine Learning

Sijing Zhong [1], Boon Kar Yap [2,3,*], Zhiming Zhong [1,4,*] and Lei Ying [1,4]

1 State Key Laboratory of Luminescent Materials and Devices, Institute of Polymer Optoelectronic Materials and Devices, South China University of Technology, Guangzhou 510640, China; msjustinzhong@hotmail.com (S.Z.); msleiying@scut.edu.cn (L.Y.)
2 Institute of Sustainable Energy, Universiti Tenaga Nasional, Jalan Ikram-Uniten, Kajang 43000, Malaysia
3 Electronics and Communication Department, College of Engineering, Universiti Tenaga Nasional, Jalan Ikram-Uniten, Kajang 43000, Malaysia
4 South China Institute of Collaborative Innovation, Dongguan 523808, China
* Correspondence: kbyap@uniten.edu.my (B.K.Y.); name.zzm@gmail.com (Z.Z.)

Abstract: Non-fullerene acceptors are promising to achieve high efficiency in organic solar cells (OSCs). Y6-based acceptors, one group of new n-type semiconductors, have triggered tremendous attention when they reported a power-conversion efficiency (PCE) of 15.7% in 2019. After that, scientists are trying to improve the efficiency in different aspects including choosing new donors, tuning Y6 structures, and device engineering. In this review, we first summarize the properties of Y6 materials and the seven critical methods modifying the Y6 structure to improve the PCEs developed in the latest three years as well as the basic principles and parameters of OSCs. Finally, the authors would share perspectives on possibilities, necessities, challenges, and potential applications for designing multifunctional organic device with desired performances via machine learning.

Keywords: Y6; non-fullerene acceptor; organic solar cell; machine learning; multi-function

Citation: Zhong, S.; Yap, B.K.; Zhong, Z.; Ying, L. Review on Y6-Based Semiconductor Materials and Their Future Development via Machine Learning. *Crystals* **2022**, *12*, 168. https://doi.org/10.3390/cryst12020168

Academic Editors: Xinping Zhang, Baoquan Sun, Fujun Zhang and Giuseppe Greco

Received: 7 December 2021
Accepted: 18 January 2022
Published: 24 January 2022

1. Introduction

With the increasing global energy demand, it is an urgent problem to study clean and renewable energy instead of traditional energy such as fossil fuels. Among the many new clean energy resources being developed, solar cells have been an area of focus due to the advantages of pollution-free, low noise, low cost of use, and no regional restrictions [1]. Presently inorganic materials [2,3] such as silicon-based semiconductors play a key role because of the high efficiency and high charge carrier mobility. However, organic solar cells (OSCs) are promising alternatives due to their multi-functional integration [4,5], superior processability [6,7], structural versatility, being light and inexpensive compared to inorganic counterparts.

Organic solar cells mainly include Schottky or single-layer solar cells, double-layer heterojunction solar cells, and bulk-heterojunction (BHJ) solar cells. The BHJ solar cell showed a significant improvement in efficiency [8]. In a BHJ solar cell, active layers containing phase separation as well as interpenetrating donor and acceptor materials are essential materials to be designed. In this review, we mainly focused on the material design of acceptors. One of the first efficient acceptors, fullerene (C_{60}), was reported by Sariciftci et al. [9]. Since then, some soluble fullerene-based acceptors such as $PC_{61}BM$ and $PC_{71}BM$ have been developed and have attracted extensive attention [10–13]. These are mainly ascribed to the unique features of the acceptors such as a large π conjugated system, a rigid molecular skeleton, and efficient photo-induced electron transfer [9,14]. A single-junction device containing $PC_{71}BM$ via hydrocarbon solvents has been fabricated, and the highest conversion efficiency has reached 11.5% [15]. However, the low efficiency compared with state-of-art organic solar cells restricts the development, which is mainly due to the materials of fullerene itself: (1) the absorption of fullerenes and their derivatives are not only weak in the spectral range but also in the extinction coefficient [16]; (2) it is hard to adjust the energy levels dramatically; (3) the spherical structure of fullerene structure is easy to crystalize and aggregate, causing long-term instability of blend morphology [17,18].

In recent years, non-fullerene small-molecule electron-acceptor (NFAs) [17,19] materials have attracted more and more attention due to adjustable energy levels, a simple synthesis process, low cost, excellent solubility, and a wider absorption range than fullerene. As a milestone, a new benchmark acceptor Y6 (2,2′-((2Z,2′Z)-((12,13-bis(2-ethylhexyl)-3,9-diundecyl-12,13-dihydro-[1,2,5]thiadiazolo[3,4-e]thieno[2,″3″:4′,5′]thieno[2′,3′:4,5]pyrrolo thieno[20,30:4,5]thieno[3,2-b]indole-2,10-diyl)bis(methanylylidene))bis(5,6-difluoro-3-oxo-2,3-dihydro-1H-indene-2,1-diylidene)) dimalononitrile)) [20] was reported in 2019 and has since been widely reported. The related reports include improving performances [21,22], understanding theories behind high performances [23], and other applications such as transparent [24,25], flexible [26], upscaling [27,28] and indoor devices [29].

In this review, the primary working mechanism and parameters for OSCs are first introduced. Secondly, the properties of Y6 materials along with the design strategy and structures are summarized. Thirdly, we summarize seven different approaches developed in the latest three years for synthesizing new materials. Finally, we focus on key challenges and some guidance suggestions on fastening material design and improving the PCE, which are mainly using machine learning as a powerful tool to screen, design, and understand the materials.

2. Basic Principle of Organic Solar Cells

Before going through the latest methods of efficiency improvement, it is necessary to understand the basic principles and parameters of OSCs. Some abbreviations shown in Table 1 and sequential molecular structures including donors, small-molecule acceptors, and solvents shown in Figure 1 are used to make the article more precise.

Table 1. Main nomenclature in this review.

OSC	Organic Solar Cell	NFA	Non-fullerene Acceptor
PCE (η)	Power-conversion Efficiency	DFT	Density Functional Theory
HOMO	Highest Occupied Molecular Orbital	BHJ	Bulk Heterojunction
LUMO	Lowest Unoccupied Molecular Orbital	TEM	Transmission Electron Microscope
GIWAXS	Grazing-incidence Wide-angle X-ray Scattering	ML	Machine Learning

A conventional solar cell of BHJ architecture with active layers containing phase separation and interpenetrating donor and acceptor materials are shown in Figure 2a. Apart from the active layer, electrodes and interlayers are stacked between the active layers. Metal oxides, salts, polymer blends, organic acids, small molecules, and conjugated compounds can be promising materials for interlayers, which serve as charge-selectivity, transport layers and can form ohmic contacts between the electrodes and active layers [30,31].

The basic principle of the organic solar cell is shown in Figure 2a, including four crucial processes: exciton formation, exciton diffusion, exciton dissociation, and charge transport and collection [32]. Under optical excitation (Process 1), electrons in the donor will be excited from the highest occupied molecular orbital (HOMO) to the lowest unoccupied molecular orbital (LUMO) and will leave holes in HOMO (Process 2). This will create an exciton or a Coulomb hole-and-electron pair. When the exciton diffuses to the interface of donor and acceptor, an exciton separates to a low-energy state. The electron transfers to the LUMO of the acceptor, and the hole remains in the HOMO of the donor, creating a charge-transfer state (CT states) (Process 3). The electrons and the holes will be collected by the electrodes (Process 4). Eventually, an electric motive force will be generated.

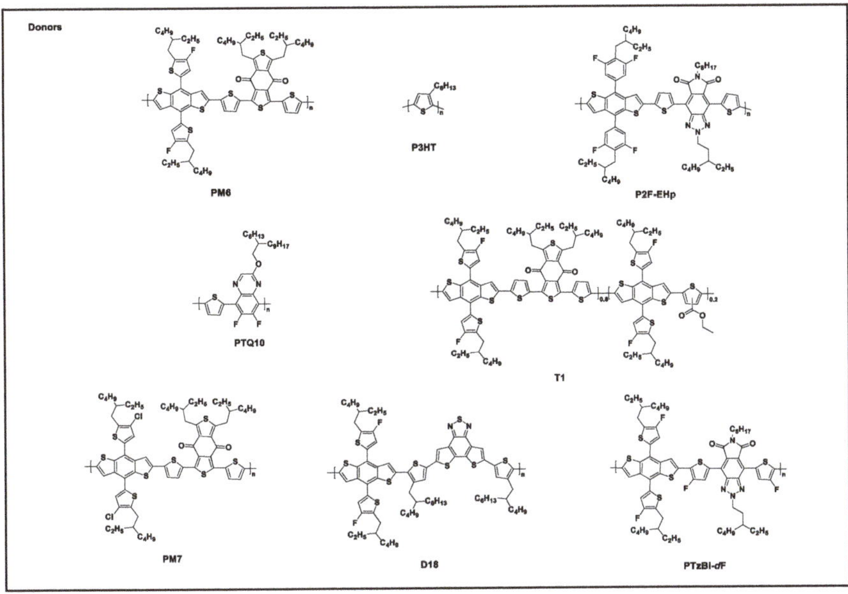

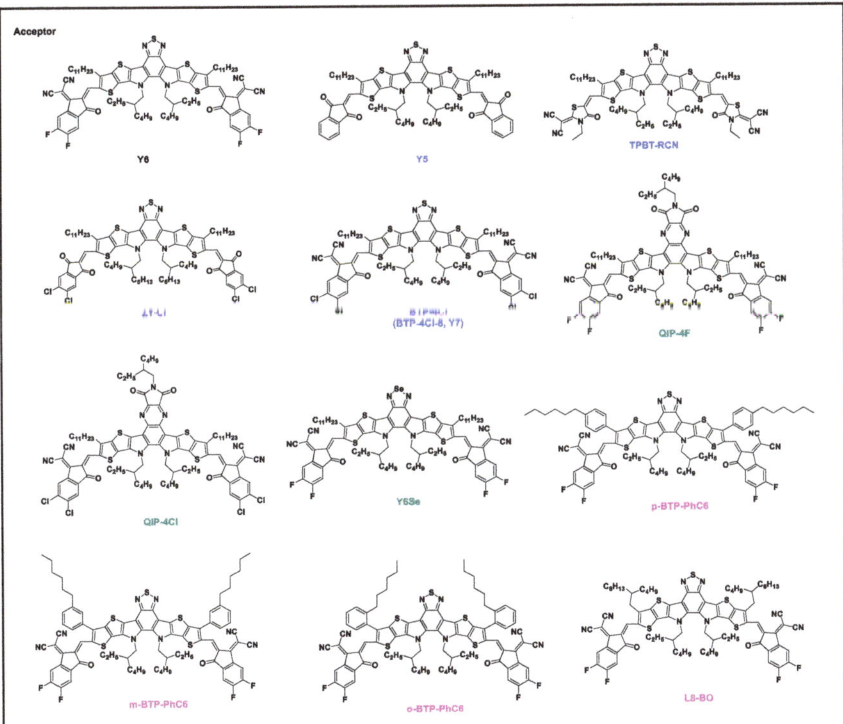

Figure 1. *Cont.*

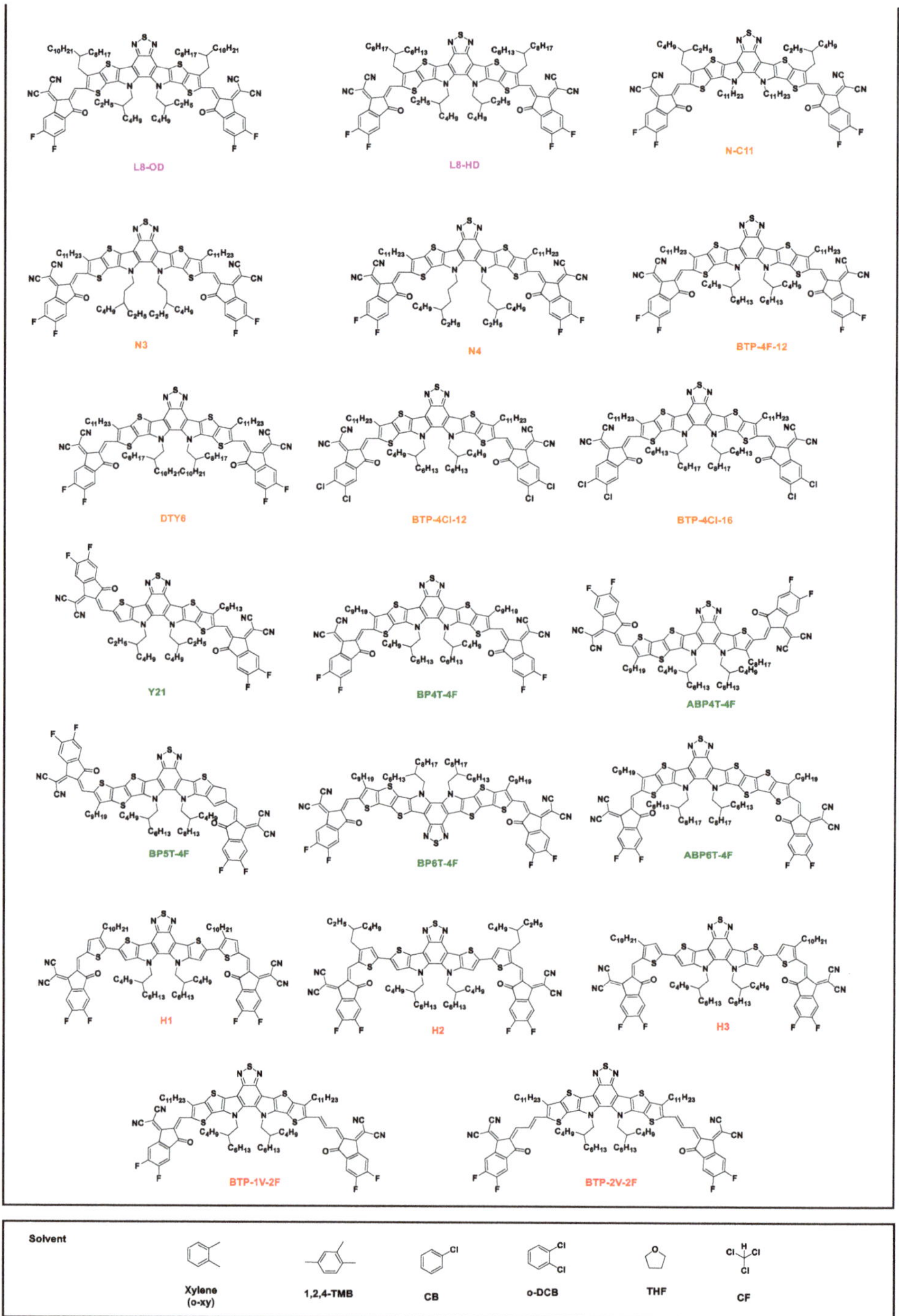

Figure 1. Main molecular structures used in the review article.

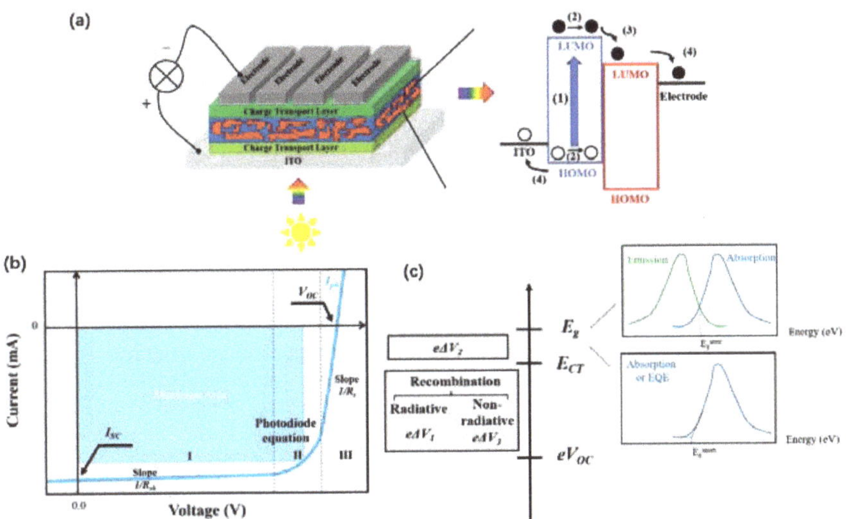

Figure 2. (**a**) The schematic structure of a bulk heterojunction (BHJ) solar cell where the blue regions represent donors and red regions represent acceptors. The enlarged picture on the right shows four basic working steps of a BHJ solar cell. (**b**) Generic *J-V* graph for the performance of a solar cell. (**c**) Different forms of energy losses with two methods calculating the bandgap shown on the right.

3. Figures of Merit of Organic Solar Cells

3.1. Power Conversion Efficiency

A typical graph demonstrating the performance is the *I-V* graph shown in Figure 2b. The most important parameters are the power-conversion efficiency (PCE or η), the ability to convert solar energy into electricity, which is defined as:

$$\eta = \frac{V_{oc} \times I_{sc} \times FF}{P_{in}}, \tag{1}$$

where V_{OC} is the voltage at an open circuit; I_{SC} is the short circuit current, i.e., the current at zero voltage; and P_{in} is the solar power absorbed, normally standardized as AM1.5 sunlight.

3.2. Fill Factor

FF is the fill factor describing the extraction of the photogenerated current [33,34], and the value can be obtained by

$$FF = \frac{max\ (V \times I)}{V_{oc} \times I_{sc}} = \frac{max\ (V \times J \times A\ )}{V_{oc} \times J_{sc} \times A}, \tag{2}$$

with the numerator being the maximum area enclosed by the curve shown in Figure 2b. *A* is the active area, and *J* is the current density, i.e., the current divided by the area through which the current flows.

Ideally, the curve will be a square shape, and thus the *FF* is 1. However, there are mainly three crucial parameters affecting *FF*, which are series resistance R_s, shunt resistance R_{sh}, and the diode [34]. The slope in region I is determined by $1/R_s$, which is mainly attributed to the resistance of the active layer, electrodes, and the interface between them. Region II is governed by a photodiode equation (active layer can be considered as pn junctions containing n-type semiconductors and p-type semiconductors):

$$I = I_s[exp(eV/nk_B\mathrm{T}) - 1]. \tag{3}$$

where I_s is the dark current; k_B is the Boltzmann constant; e is the elementary charge; T is the temperature; and n is a correction factor controlled by recombination and dissociation. The sharp slope in region III is dominated by $1/R_{sh}$, which exists due to the current leakage from the pinhole and edge in the device.

3.3. Short Circuit Current Density

J_{SC} was found to be the dominant factor of the PCE [35]. Photons absorbed, and free charge carriers generated and collected determine J_{SC} [35]. Technically, the smaller the bandgap, the more overlap between the absorption of the active layer and the solar spectrum, resulting in the more potential to achieve a higher J_{SC}. However, a smaller bandgap could also result in a lower V_{OC} in terms of the same voltage loss. Li and coworkers [36] have found that an ideal NFA with a bandgap of about 1.4 eV could reach 19.8%.

3.4. Voltage Loss

Voltage loss (V_{loss}) originates from three losses [37,38] shown in Figure 2c and can be written as

$$V_{loss} = E_g/e - V_{OC} = \Delta V_1 + \Delta V_2 + \Delta V_3, \tag{4}$$

where ΔV_1 represents the intrinsic loss caused by radiative recombination above the bandgap; ΔV_2 denotes the additional radiative recombination below the bandgap; ΔV_3 stands for the loss caused by non-radiative recombination; and E_g can be determined by the absorption onset $\left(E_g^{\text{onset}}\right)$ and a more accurate method determined by the interception of the absorption and emission spectrum $\left(E_g^{\text{inter}}\right)$.

We can see that V_{OC} can reach E_g/e when there is no loss. ΔV_1 is unavoidable, and it is typically between 0.25 and 0.3 V [38]. So, we can control the other losses by high illumination efficiencies and an aligned band structure (a smaller energy difference between two different LUMO) [38]. For example, the losses can be significantly reduced from around 0.9 eV to about 0.5 eV when replacing a fullerene acceptor with an NFA [17,23].

4. Synthesis and Properties of Y6 Material

To simplify, all the synthesis processes and properties of Y6 are cited from Yuan et al [20].

4.1. Material Design

Y6 material (Figure 3a), also known as BTP-4F [21], BTF-4F-8 [39], and BTPTT-4F [40], is an n-type organic semiconductor or non-fullerene acceptor. The molecule design synthesis was based on a recently used push–pull strategy [41–44], A-DAD-A, where "A" represents the accepting electrons or the electron-withdrawing moiety in green regions in Figure 3a, and "D" denotes the donating electrons or electron-donating moiety in blue regions in Figure 3a. Firstly, commercially available 2, 1, 3-benzothiadiazole (BT) [45,46] was employed as the central core, which not only serves as constructing low-bandgap materials [8] but can also be a potential candidate in thick-film devices [47] due to the high hole mobility of the BT unit [48]. Secondly, based on the BT unit, a novel dithienothiophen[3.2-b]-pyrrolobenzothiadiazole (TPBT) unit was introduced to extend the conjugation length. Thirdly, long linear and branched alkyl chains in "D" units were used to possess high solubility and the opportunity of preferable packing, which will be later discussed in "progress of the improvement of Y6 material" section. Finally, 2-(5,6-difluoro-3-oxo-2,3-dihydro-1H-inden-1-ylidene)malononitrile (2FIC) as end units were encapsulated, which are believed to promote intermolecular interactions by non-covalent bondings [49,50] and to enhance optical absorption [20]. Zhang and coworkers [23] reported that delocalization of exciton and electron wave functions in the Y6 material and distinctive two-dimensional

$\pi - \pi$ molecular packing in solution and thin films illustrate a good material design for highly efficient organic solar cells.

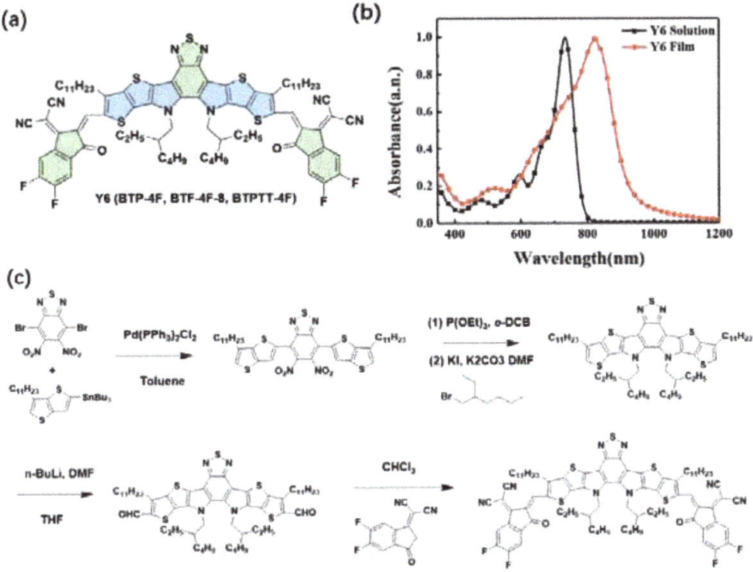

Figure 3. (**a**) Structure of Y6 material where the green regions represent electron-withdrawing unit or acceptor unit, and the blue regions stand for electron-donating unit or donor unit. (**b**) UV–Vis–NIR absorption spectra of the thin film and solution for Y6 [20]. Reprinted with permission from Ref. [20]. Copyright 2019 Elsevier Inc. (**c**) Synthesis of Y6 material with yield rates in each step [20]. Adapted with permission from Ref. [20]. Copyright 2019 Elsevier Inc.

4.2. Material Synthesis

Dark-blue Y6 powder was synthesized with only four steps and an overall yield of 12% [20]. The raw materials and the famous name reactions including Stile coupling, Cadogan reductive cyclization, the Vilsmeier–Haack reaction, and Knoevenagel condensation for the whole process are shown in Figure 3c.

4.3. Absorption Behaviors

The absorption spectra of Y6 in the solution (black line) and film (red line) are shown in Figure 3b. Y6 exhibits strong and broad absorption in the 600–1000 nm region and extends to 1100 nm corresponding to the near-infrared region with a maximum absorption peak at 821 nm and an absorption coefficient of 1.07×10^5 cm^{-1}. The absorption coefficient of Y6 material is higher and red-shifted compared with another common NFA material, ITIC [19]. Moreover, low absorption in the range of 400 nm to 600 nm can be observed in solution and the thin film of Y6, indicating potential applications for transparent optoelectronic devices. The optical bandgap (E_g) of Y6 is 1.33 eV (close to 1.4 eV), calculated from $E_g = \frac{1240}{\lambda_{onset}}$ where the absorption onset (λ_{onset}) of Y6 is 931 nm.

The absorption range of Y6 material is complemented with that of PM6 materials where its absorption range is within 400 and 600 nm. The two materials make full use of the photons in the solar-radiation spectrum, making it one of the main reasons for high J_{SC} in OSCs.

4.4. Band Properties

To evaluate the practical HOMO and LUMO, the band properties of Y6 were measured via electrochemical cyclic voltammetry. They used anhydrous acetonitrile (CH_3CN) solution with Ag/AgCl as a reference electrode and the ferrocene/ferrocenium (Fc/Fc^+) as an internal reference. The energy levels of HOMO and LUMO for Y6 are −5.65 eV and −4.10 eV, respectively [20].

The voltage loss of the PM6:Y6-based device is 0.67 V by E_g^{onset}, which was much smaller than fullerene and some non-fullerene devices with a smaller bandgap at that time [23,38], where PM6 is also known as PBDB-TF [51], PBDB-T-F [52], or PBDB-T-2F [23].

Other properties such as sufficient thermal stability, high electron and hole mobilities, and good morphology properties can also be found in Ref. [20]. In conclusion, compared with ITIC-series- [53] and M-series- [54] based solar cells, higher utilization of sunlight, lower energy loss, and a high fill factor give rise to high efficiency for conventional and inverted structure [55,56] of the device, both demonstrating the high efficiency of up to 15.7% [20].

5. Progress of the Improvement in Y6 Material

After successful synthesis of the Y6 NFA, scientists have tried different methods to improve the material including new acceptors retaining Y6′s main properties; new donors matching Y6 including P2F-Ehp [40], PtzBI-dF [57], PM7 [58], and D18 [59]; and device engineering such as the ternary strategy [60–62], the quaternary strategy [63], and layer-by-layer structures [64]. For Y6 derivatives, there are mainly seven different material engineering methods shown in Figure 4, and the following discussion will be based on this figure.

5.1. Fine-Tuning the Flanking Unit

Altering the end group provides an effective method to tune the energy levels shown in Figure 5a.

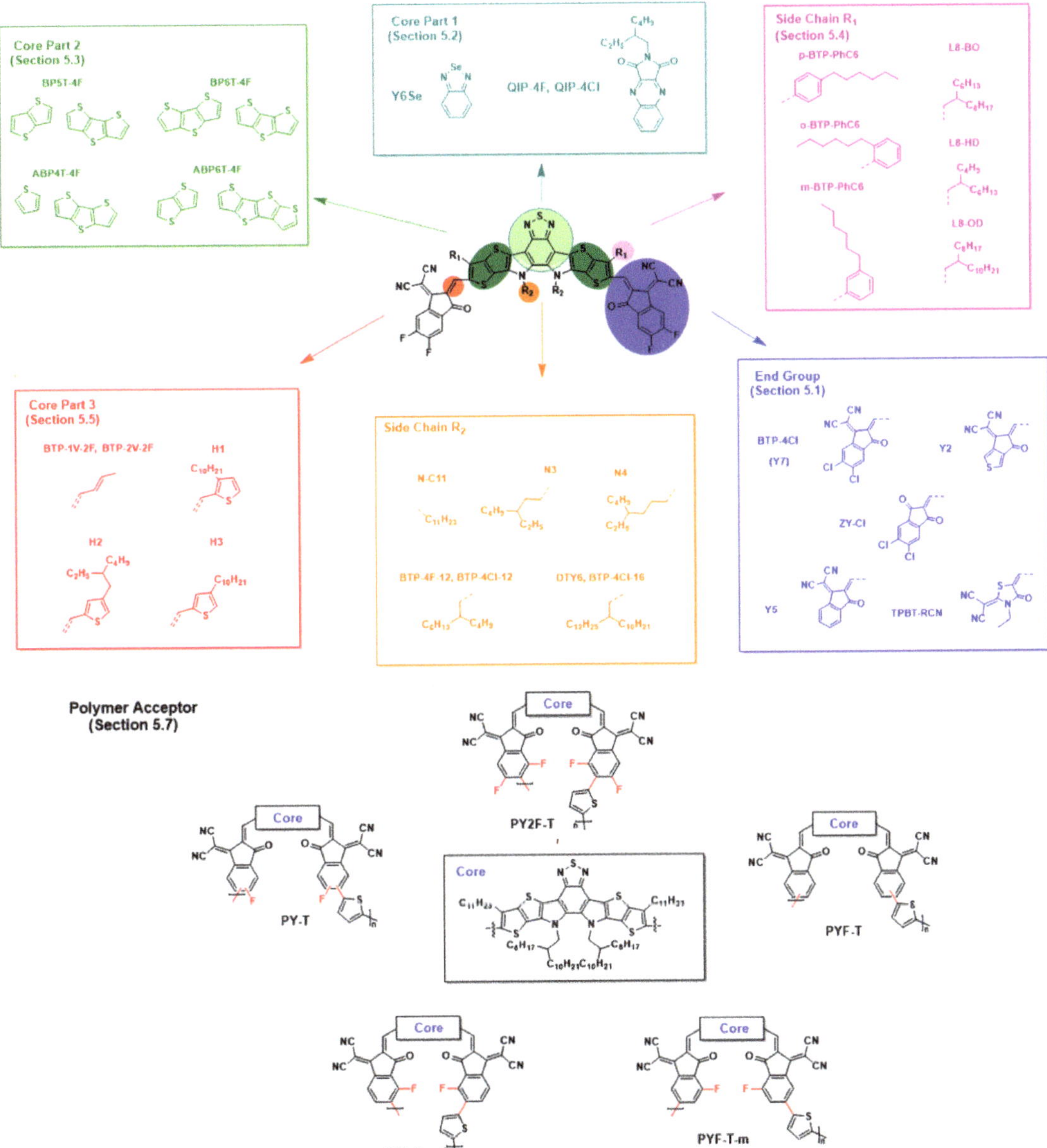

Figure 4. Seven main approaches to alter Y6 to improve efficiency.

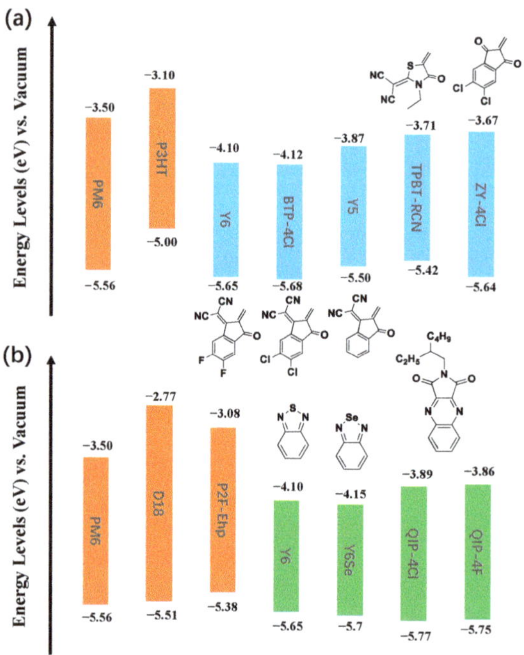

Figure 5. Frontier molecular orbits when (**a**) altering flanking units and (**b**) nucleus in the middle. Data from Refs. [20,21,65–68].

P3HT is the simplest and cheapest donor that can more possibly be industrialized [69]. However, in a P3HT:Y6 system, the significantly large difference in LUMO and HOMO of the active layers gives rise to high voltage loss (0.86 V), low V_{OC} (0.45 V), and consequently a low efficiency (2.41%) [65]. Using a weaker non-fluorine electron-withdrawing group to encapsulate the core, Y5 shows better-aligned energy levels with P3HT, leading to a PCE of 3.60% [65]. Yang and coworkers [65], in 2020, reported a new TPBT-RCN acceptor modifying the end-capping groups of Y6 via an even weaker group. The absorption onset of the new NFA acceptor, TPBT-RCN, with a weaker end group leads to a significant blue-shift to that of Y5 and Y6. However, better-matched energy levels, more photon absorbing in the visible spectrum and higher mobility could be attributed to a comparable J_{SC} of 16.49 mA cm^{-2}, a higher V_{OC} of 0.81 V, and an efficiency of 5.11% [65].

Hou and his group [66] in 2020 synthesized novel ZY-4Cl n-type semiconductors with another end group enabling a different band structure and better morphology of nanoscale phase separation. Consequently, *Jsc* and *FF* improve significantly to 16.49 mA cm^{-2} and 0.65, respectively, with a little decline in V_{OC} in comparison with that of TPBT-RCN. Finally, a P3HT:ZY-Cl device achieves 9.5%, which was the highest reported efficiency of P3HT-based OSCs at that time [66].

In 2019, Cui and coworkers [21] changed the F-substituted to the Cl-substituted terminal unit of the Y6 and reported the acceptor BTP-4Cl that is also later known as Y7 [26]. Usually, chlorination results in a downshifted LUMO level and thus a reduced V_{OC} in the device [70,71]. However, an abnormal phenomenon was observed. BTP-4Cl (Y7) materials show a red-shifted absorption compared with BTP-4F (Y6), but they also demonstrate increased V_{OC} due to the V_{loss} decreasing by 0.04 eV. These properties together give a higher efficiency of 16.5% [21], matching with the PM6 donor.

5.2. Creating a New Core Moiety

Synthesizing a new center core is another method to change the electronic properties, especially the band properties shown in Figure 5b.

Zhu and coworkers [68] in 2020 employed a novel imide-functionalized quinoxaline (QI) in the central unit and synthesis of QIP-4Cl and QIP-4F, resulting in a deeper HOMO and a higher LUMO level shown in Figure 5b. These achieve an impressive high V_{OC} of 0.94 V and thus the highest efficiency of 13.3% [68] for P2F-Ehp:QIP-4Cl. This efficiency wass among the best QI-based binary device at that time. These materials also suggest that the end group can alter the band structure too.

In late 2020, Zhang and coworkers [67] have reported Y6Se by substituting selenium for Sulphur in the central building block. This strategy has successfully decreased the radiative (ΔV_2) and non-radiative (ΔV_3) recombination loss as well as a little broader absorption, higher mobility, and better photostability. All of these show the highest efficiency of 17.7% [67] at that time.

5.3. Altering the Donor Units in A-DAD-A

Modifying the donor units in A-DAD-A, especially the emerging asymmetric molecule design, is an effective method to control the molecule stacking.

In 2020, Cai and coworkers [72] first unidirectionally removed thiophene in the ladder benzothiophene unit, resulting in an asymmetric molecular structure, Y21, and a larger dipole moment than that of Y6. This larger dipole moment would reinforce the molecular packing, and, consequently, the efficiency of the PM6:Y21 sample is 15.4% [72].

In 2021, Gao and coworkers [73] laterally fused four and five thiophene units into benzothiadiazole to produce BP4T-4F and BP5T-4F. Besides, a highly asymmetric BP4T-4F molecule (ABP4T-4F) with one thiophene on one side and three thiophenes on the other side were also synthesized and studied. From Figure 6a–c, the signal of BP4T-4F is the most distinct, meaning the favorable face-on packing, whereas that of ABPT-4F is the least. Blended with PM6, the PCEs of BP4T-4F, BP5T-4F, and ABPT-4F are 17.1% [73], 16.7% [73], and 15.2% [73], respectively. Gao and coworkers [74] also tried to fuse six thiophene units into the Y6 core and constructed a new acceptor BP6T-4F. In addition, another isomer molecule of BP6T-4F, namely, ABP6T-4F, was synthesized based on the isomerization of asymmetric strategy. However, this leads to quite opposite effects in the efficiency. In Figure 6d,e, the diffraction signal in ABP6T-4F shows a stronger signal in the q_r direction (stronger π-π stacking property) even though the signal is weaker in the q_z direction (weaker crystallization property). The advantages of the symmetry-breaking strategy in this molecule far outweigh the disadvantages, thus resulting in a significantly improved efficiency of ABP6T-4F (15.8% [74]) than that of BP6T-4F (6.4% [74]).

5.4. Tuning the Alkyl Chain R_1

Sidechain modification on the beta position of the thiophene unit, R_1, is also a flexible strategy mainly used to control intermolecular stacking.

Chai and coworkers [75] focused on the side-chain orientation by 2021. They designed and synthesized three isomeric NFA named o-BTP-PhC6, m-BTP-PhC6, and p-BTP-PhC6. The results show that the hexyl chain of m-BTP-PhC6 is titled in terms of the molecule plane in comparison with vertical o-BTP-PhC6 and the horizontal p-BTP-PhC6 shown in Figure 7a by DFT (density functional theory) calculation where the 2-ethylhexyl chains attached to the pyrrole rings were replaced by 2-methyl propyl to accelerate the calculations. Due to this unique orientation, m-BTP-PhC6 exhibits the most advantageous intermolecular stacking among the three isomers. Figure 7b shows that the m-BTP-PhC6 peak is the most prominent in the q_z direction between 1.5 A^{-1} and 2 A^{-1}, which indicates a face-on direction beneficial to light-harvesting where q_z is the magnitude of the scattering vector in the z-direction. When the molecule is mixed with PTQ10, the PCE of the device based on m-BTP-PhC6 reaches 17.7% [75], which is significantly better than that of the device based on o-BTP-PhC6 (16.0% [75]) and p-BTP-PhC6 (17.1% [75]) and the best value of non-fullerene OSC device based on PTQ10 at that time.

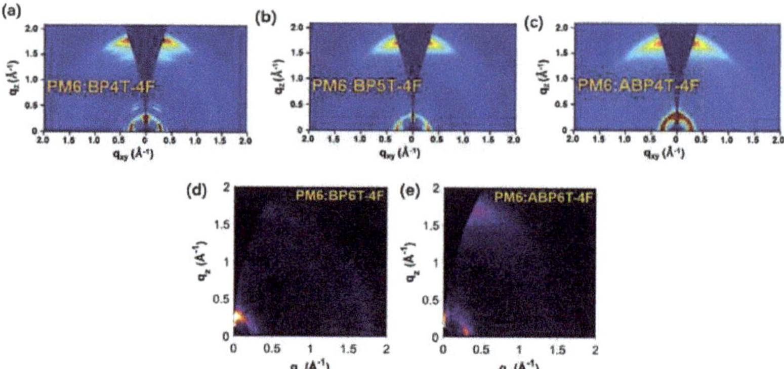

Figure 6. 2D-GIWAXS patterns of PM6 blended with (**a**) BP4T-4F [73] (adapted with permission from Ref. [73]. Copyright 2020 WILEY-VCH GmbH.), (**b**) BP5T-4F [73] (adapted with permission from Ref. [73]. Copyright 2020 WILEY-VCH GmbH.), (**c**) ABP4T-4F [73] (adapted with permission from Ref. [75]. Copyright 2020 WILEY-VCH GmbH.), (**d**) BP6T-4F [74] (adapted with permission from Ref. [74]. Copyright 2021 WILEY-VCH GmbH.), and (**e**) ABP6T-4F [74] (adapted with permission from Ref. [74]. Copyright 2021 WILEY-VCH GmbH.).

Li and coworkers [22] by 2021 have created L8-R (L8-BO, L8-HD, and L8-OD) acceptors through a branched-chain in R_1. Compared with the two kinds of Y6 molecule π–π stacking, there are three stacking patterns for L8-R molecules shown in Figure 7c through the Zerner's intermediate neglect of differential overlap (ZINDO) method [76], which can provide more charge-hopping channels, which are thus more favorable for the transition of charge. Among the three NFAs, the electron mobility of L8-BO (2-butyl octyl substitution) was 6.79×10^{-4} cm^2 V^{-1} s^{-1}, which is higher than the mobilities of Y6 (4.49×10^{-4} cm^2 V^{-1} s^{-1}), L8-OD (4.87×10^{-4} cm^2 V^{-1} s^{-1}), and L8-HD (5.54×10^{-4} cm^2 V^{-1} s^{-1}) [22]. Besides, TEM images in Figure 7d show that L8-BO has the best morphology due to its uniformity and no obvious aggregation to impede charge transfer. These factors together improve the performance of the device. When L8-BO is blended with PM6, the efficiency of the single junction photovoltaic device is 18.32% [22].

These results show that Y6 can still have more room to be improved, reaching 20% via different methods.

5.5. Extending the Conjugation Length

It is useful to extend the conjugation length to absorb more photons in the near-infrared region. The absorption and photovoltaic properties of Y6 and some tailored red-shifted materials are shown in Table 2.

He and coworkers [77], by 2020, successfully employed different types of alkyl thiophenes to extend the absorption onset of Y6. As seen from Table 2, H1, H2, and H3 bathochromic-shift 1016 nm. Interestingly, the different alkyl chains in the thiophenes again demonstrate different packing, namely, edge-on for H1, mixed with edge-on and face-on for H2, and dominantly face-on for H3. Consequently, H3 exhibits the highest efficiency of 13.75% among the three new materials. Li and coworkers [24] later employed the H3 to semi-transparent organic solar cells and achieved the PCE of 8.26% [24] and average photopic transmittance (APT) of 47.72% [24].

Hai and coworkers [78] attached vinylene to one side (BTP-1V-1F) and both sides (BTP-1V-1F) to the core of the Y6. Although BTP-2V-2F red-shift not as obviously as H3 and BTP-2V-2F (Table 2), its efficiency (14.24% [78]) is one of the highest values for binary single-junction OSCs based on ultra-narrow bandgap NFA with a bandgap below 1.29 eV [78].

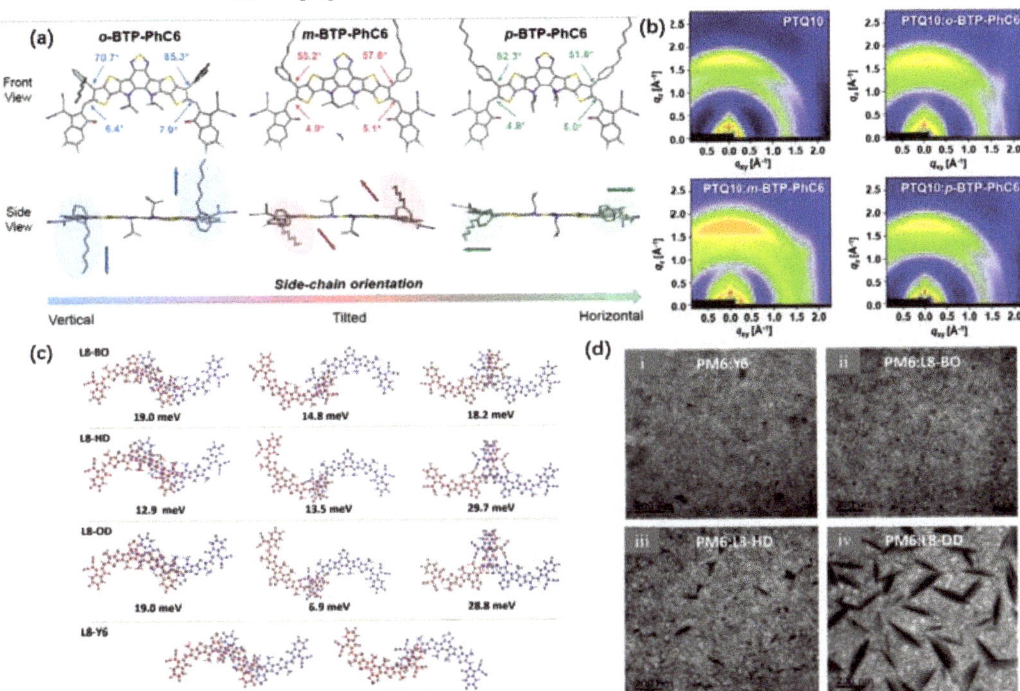

Figure 7. (**a**) Optimized geometry o-BTP-PhC6, m-BTP-PhC6, and p-BTP-PhC6 of by DFT calculation [75]. (**b**) 2D GIWAXS images of a neat PTQ10 film; PTQ10:o-BTP-PhC6, PTQ10:mBTP-PhC6, and PTQ10:p-BTP-PhC6 blend films [75]. (**c**) Electronic coupling (meV) for L8-R (L8-BO, L8-HD, L8-OD), and Y6 dimers with different crystal packing motif. The electronic couplings for dimers of L8-R and Y6 in crystals were computed [22]. Adapted with permission from Ref. [22]. Copyright 2021 Springer Nature Limited. (**d**) TEM patterns of PM6:Y6 and PM6:L8-R blend films [22]. Adapted with permission from Ref. [22]. Copyright 2021 Springer Nature Limited.

Table 2. Absorption and photovoltaic properties of Y6, H1, H2, H3, BTP-1V-2F, and BTP-2V-2F [78]. Data from Refs. [20,77,78].

Acceptors	λ_{onset}^{film} (a) (nm)	Matched Donor	J_{SC} (mA cm^{-2})	V_{OC} (V)	*FF* (%)	PCE (%)
Y6	931	PM6	25.3	0.83	74.8	15.7
H1	1016	PBDB-T	16.81	0.784	53	6.98
H2	1016	PBDB-T	24.40	0.781	69	13.15
H3	1016	PBDB-T	25.84	0.757	70	13.75
BTP-1V-2F	984	PM6	24.75	0.80	72	14.24
BTP-2V-2F	1020	PCE10	26.50	0.66	64	11.22

(a) Cast from chloroform solution.

5.6. Modulating the Side Chain R_2

Tailoring the side chain in the pyrrole unit, R_2, could be mostly used to increase solubility and apply to a large-scale device while maintaining nearly the same absorption and band profiles.

Jiang and coworkers [79] have first synthesized N-C11 by swapping the R_1 and R_2 chains of Y6. They found that N-C11 demonstrates much lower solubility and larger domain (56.4 nm compared with 21 nm in Y6 measured by resonant soft X-ray scattering or RsoXS in short), which impairs the efficiency of the device. In light of this, it is vital to keep the branched alkyl chain on the nitrogen atom of the molecule to retain steric hindrance, and the branched position of the branched alkyl chain on the pyrrole unit is further optimized. The group also synthesized N3 and N4 via shifting branching units of R_2. Comparing three kinds of molecules (Y6, N3, and N4) with different branched-chain positions, it was found that the acceptor with the third branched alkyl chain has the best solubility, morphology (Figure 8a), and electronic properties, thus achieving the best efficiency of 16.0% [79]. Finally, a highly efficient organic solar cell containing the PM6 with a PCE of 16.74% [79] was obtained by using the ternary strategy when adding a small amount of $PC_{71}BM$ receptor.

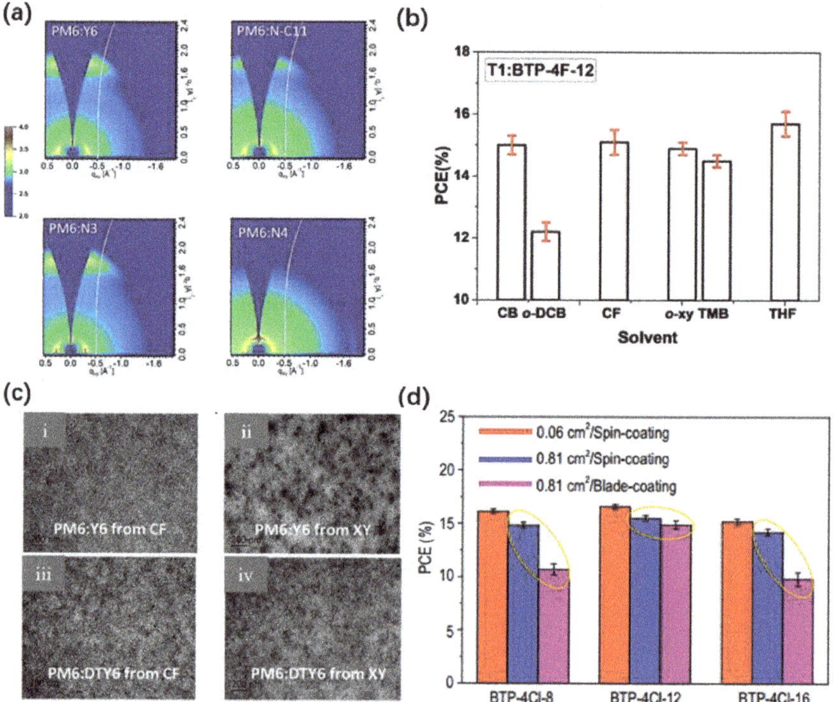

Figure 8. (**a**) 2D GIWAXS images of PM6:Y6, PM6:N-C11, PM6:N3, and PM6:N4 blend films [79]. Adapted with permission from Ref. [79]. Copyright 2019 Elsevier Inc. (**b**) Statistics of PCEs for OSCs based on T1:BTP-4F-12 with different solvents [39]. Adapted with permission from Ref. [39]. Copyright 2019 WILEY-VCH Verlag GmbH & Co. KGaA, Weinheim. (**c**) TEM images of PM6:Y6and PM6:DTY6 blend films cast from two solvents (chloroform and o-xylene) [7]. Adapted with permission from Ref. [7]. Copyright 2020 Elsevier Inc. (**d**) Histogram of PCEs for BTP-4Cl-X-based OSCs under different conditions, where 0.06 cm^2 and 0.81 cm^2 are mask areas with active areas or device areas of 0.09 and 1 cm^2, respectively [80].

Hong and coworkers [39] in 2019 reported that a novel small-molecule acceptor material BTP-4F-12 had been synthesized successfully by increasing the length of the branched alkyl chain R_2 (from 8 carbons to 12 carbons) on the Y6 non-fullerene acceptor. This strategy shows improved efficiency of the single-junction binary OSC to 16.4% [39] with PM6 due to the improvement of crystallinity and electron mobility. Further studies had shown that replacing the donor T1 with even better solubility enables the possibility of different kinds of eco-compatible solvents to fabricate OSCs, with the results shown in Figure 8b. More importantly, with tetrahydrofuran (THF), a non-chlorinated and nonaromatic solvent, a high PCE of 14.4% [39] was achieved with an active layer of 1.07 cm^2 by blade-coating rather than traditional spin-coating [69].

Dong and coworkers [7] by 2020 also tried to extend the side chains and have reported a non-fullerene acceptor DTY6. When mixed with donor PM6, the OSCs based on DTY6 show an excellent PCE of 16.3% [7] when using non-halogen solvent o-xylene (XY), whereas Y6-based OSC showed poor device performance (PCE < 11% [7]) when treated with XY. Due to the aggregation of Y6 in XY (Figure 8c), very large domains appear in Y6-based hybrid films, which leads to the low efficiency of hole transfer from Y6 to PM6 and enhances the non-radiative recombination to 0.28 eV [7]. However, the films based on DTY6 display a more reasonable domain size (Figure 8c), which ensures effective hole transfer and low non-radiative recombination (0.24 eV [7]) from DTY6 to PM6. A large active layer area of 18 cm^2 with opaque PM6: DTY6-based components treated with o-xylene were fabricated via the blade-coating method with a PCE of 14.4% [7].

Cui and coworkers [80] in 2019 applied this strategy to Y7 materials and synthesized a series of NFA BTP-4Cl-X (X = 8, 12, or 16). They also blended the BTP-4Cl-X acceptor materials with the polymer-donor material PM6 and prepared OSCs by a spin-coating method as well as a blade-coating method with different areas of the active layer, respectively. The results in Figure 8d show that BTP-4Cl-12 achieves the highest efficiency in the two processing methods with different areas, indicating that the most preferable processability and thus the most suitable morphology characteristics. More specifically, small cells with an active area of 0.09 cm^2 (0.06 cm^2 mask area) by spin coating and larger cells with an active area of 0.81 cm^2 (1 cm^2 mask area) prepared by blade coating achieved a PCE of 17% and 15.5%, respectively [80].

5.7. Polymer Acceptor

All-polymer solar cells (all-PSCs) composed of a mixture of a polymer donor and a polymer acceptor have attracted much attention due to better flexibility, easier large-scale production, and higher transparency [81–85]. However, the PCE of the pure polymer blend system is still far behind that based on the polymer donors and small-molecule acceptors' counterparts. This is mainly due to the lack of high-performance polymer acceptors including a high absorption coefficient and a low energy bandgap and the poor mutual solubility with polymer donors due to the self-aggregation of polymers [86]. In this section, we discuss some tailored polymers structures (Figure 4) based on Y6.

Wang and coworkers [86] have successfully synthesized a series of PYT polymers based on the Y6 structure with low, medium, and high molecular weights, namely, PYT_L, PYT_M, and PYT_H, all of which exhibit a low optical bandgap of 1.40–1.44 eV and a high absorption coefficient of over 10^5 cm^{-1}. The relationship between molecular weight and various physical properties was also studied. PYT_M has better miscibility, higher mobility, and less loss. Therefore, the final device with PYT_M:PM6 shows a 13.44% high efficiency, better than that based on PYT_L (12.55%) and PYT_H (8.61%) [86].

Compared with the non-fluorinated compound PYT, the PYF-T reported by Yu and coworkers [87] in 2021 has a stronger and red-shifted absorption spectrum, stronger molecular packing, and higher electron mobility. At the same time, the fluorination on the terminal group of PYF-T leads to the downshift of its energy level and the higher matching with the donor PM6, allowing a higher charge transfer and a smaller voltage loss. Therefore, the device efficiency of PM6: PYF-T is 14.1% [87]. A similar strategy was applied to another polymer acceptor PFA1 by Peng and coworkers [88] in 2020, with a dramatically improved efficiency of 15.11% paired with PTzBi-Si, where the non-fluorinated counterpart only results in a PCE of 4.01% [88].

Two regionally specific polymer acceptors named PYT-T-o and PYF-T-m had been prepared by Yu and coworkers [89] by 2021. Compared with the random polymer PYF-T and PYF-T-m with weaker conjugation and the intramolecular charge transfer effect, PYF-T-o shows a more regular molecular arrangement, stronger and red-shifted absorption, and more ideal phase separation, all of which lead to the device efficiency of PM6:PYF-T-o to 15.2% [89].

Based on PYF-T, Yu and coworkers [90] in 2021 reported a PY2F-T with two fluorine atoms on its one end group, and an all-polymer solar cell with an efficiency of 15.22% [90] was obtained. After introducing the fluorine-free PYT as the third component into the main system of PM6: PY2F-T by Sun and coworkers [91], the improved PCE of PM6:PY2F-T is as high as 17.2% [91] due to a balance between material crystallization and phase separation.

6. Future and Perspectives

Exploring ever more powerful donor or acceptor materials should be the top priority to achieving the high efficiency of OSCs. The design of organic materials is far more fundamental and complex than the device engineering issues. In terms of material development, continuing to design and modify new materials is important, with the emphasis on ML- [92] guided molecular design.

6.1. Possibilities

Since the first report of OSCs in 1973 [93], around 2000 donor molecules in use of OSCs have been synthesized and tested in photovoltaic cells [94]. These donors provide quite efficient data for high-dimension and/or deep ML. Additionally, the computing speed of the computer and ML have developed rapidly, especially the prevalence of deep learning and graphics processing units (GPU). Sun and coworkers [94] by 2019 established a dataset containing 1719 donor materials and used a variety of ML algorithms for binary classification. In the end, the best ML algorithm was independently validated by synthesizing 10 new OSC donor materials. The accuracy of the 10 new materials is 80% [94], which is in good agreement with the experimental classification results. These together give the possibilities to use ML to assist molecule design.

6.2. Necessities

Employing ML is not only possible but also desirable and necessary. As discussed in the review previously, from fullerene-based acceptors to non-fullerene acceptors or from the successful trial of the Y6 material, scientists always do a large number of trials and errors in different aspects such as tuning the side chains, trying new flanking units, and designing new building blocks. It is fine and admirable since these experiments provide a huge amount of data and experience, which is of great use for further development. However, the experience and accuracy of human beings are often inferior to those of computers. It is important to note that trials fail, and the successful reports seem inspiring but few. This is especially true that at least one property would be affected when other properties are being improved. For example, when improving J_{SC}, the two parameters—V_{OC} and *FF*—would usually reduce.

6.3. Future of Machine Learning in OSCs

The main process of combining ML with traditional design is shown in Figure 9. Datasets, including different predicted values (*Y*s) and the relevant factors or features (*X*s) such as molecular structure (*M.S*) and processing conditions (*P.C)*, need to be built. Computers, which can only identify numbers, cannot recognize molecular structures like human beings. Therefore, molecular structures as well as other relevant parameters need to be converted to numbers by feature extractors. *X*s and *Y*s would be then put into the machine, which can learn a model or a function via ML. After training the model to have the lowest losses or highest accuracies, the molecular design could be obtained quickly.

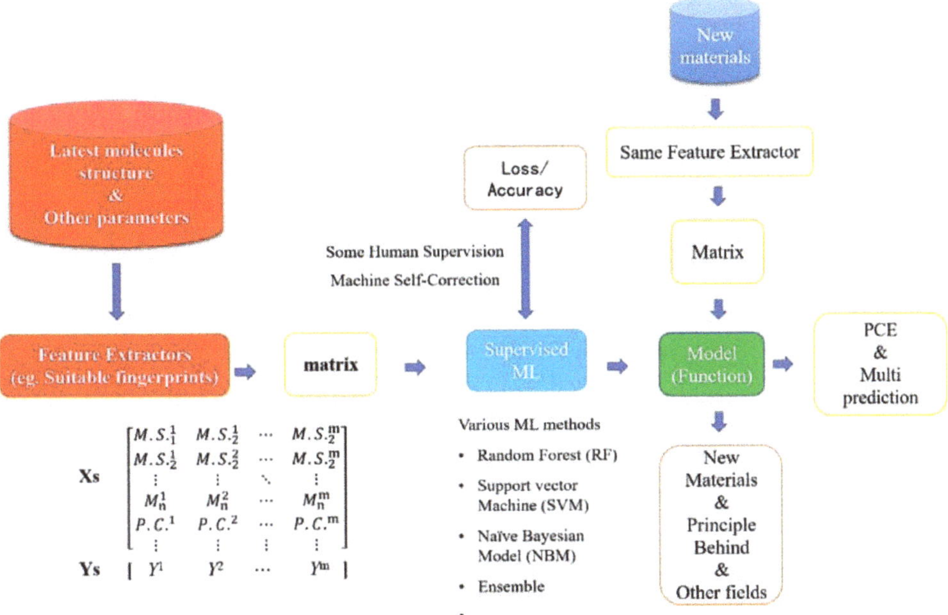

Figure 9. Possible directions and strategies in machine learning in OSCs where *M.S.* refers to molecular structure; M_n refers to number-average molecule weight; *P.C.* refers to processing condition; and superscript indicates the index of data. Red boxes indicate the most important areas for material scientists.

The tools for ML workflow are available and are user-friendly for material scientists. The key issue is setting up a high-quality database and features extractor for a given material. Although scientists have carried out substantial research [35,94–102] on ML for OSCs, there are still problems that will be discussed in Section 6.3.1. More applications will be discussed in Section 6.3.2.

6.3.1. Key Challenges

(a) New and more Datasets

Some reports [35,98–100] on machine learning are based on fullerene-based single-junction OSCs, which have limited efficiency, and fewer people have focused on it compared with NFAs. The dataset of 1719 OSC donor materials set up by Sun and coworkers [94] includes both new and old materials. However, these data contain some old-reported ones, which are probably meaningless and even harmful for ML. This is because old-reported ones will have much higher efficiency now due to the accumulation of experience and significantly improved device-engineering strategies. So, in order to achieve high accuracy of ML, we must first establish high-quality data. Some efficient methods include web crawl and a Natural Language Process (NLP) [103], which could be further established. Additionally, building some online platforms in public to renew, check, and train the data is an effective method, which could probably act as a magnet to attract the research attention. In addition, more predicted values including efficiency, transparency, *FF*, stability, processability parameters, and thermal properties could be concluded in the dataset.

(b) Suitable Fingerprints

Most feature extractors on ML for organic materials mainly come from fingerprints. There are more than 10 kinds of fingerprints with more than 15 software applications that could convert molecular structures to corresponding fingerprints [104]. Fingerprints, containing many bits ranging from 166 (MACCS) [105] to 13,824 (TGT) [106], are mature now especially in drug discoveries [107,108]. However, these fingerprints are not quite suitable for ML in OSCs. Firstly, the more bits the fingerprints contain, the more information they convey; but they are more time-consuming when performing ML. Besides, those fingerprints are mainly used in pharmaceutics, of which some parts are trivial for pharmaceutics but may be significant for the design of OSCs due to totally different molecular design purposes. Secondly, increasing one dimension means significantly more data needed to be input [109]. Even with 166 bits and containing nearly all the useful information, these 166 dimensions for the input system would be a disaster when there are at most 1719 relatively low-quality data available.

6.3.2. Applications

Possible applications mainly include the following part, which is sorted mainly according to the difficulty of implementation.

(a) Filtering before Synthesis

Filtering a material before synthesis may be the easiest part. Once we want to synthesize a new material, we can put it into the trained model to see whether it performs high efficiency before conducting the experiment. If the value is not high, iterations could be carried out until an acceptable value is achieved. This will reduce some unnecessary costs for designing low-efficiency materials. Note that the algorithm should allow the ceiling efficiency to be high instead of low when achieving the same accuracy. Besides, we can also allow predicted values such as mobility and absorption onset, so that materials used in other areas such as conducting polymers would be guided.

Based on the functions predicted, required molecules are known, and even the highest efficiency or the highest mobilities could be derived. Take the three-dimensional input as an example: we can solve for the maximum value of the function and subsequently get an exact 3-bits fingerprint, which can, in turn, get back the corresponding molecule.

(b) Predicting and even Unifying More Things Together

After synthesizing or choosing materials, processing conditions could also be optimized via machine learning. Du and coworkers [102] by 2021 have combined robot and ML to screen processing conditions in terms of efficiency and photostability. Furthermore, we can combine material properties as well as different processing conditions to design materials or devices by extrapolation if we could set up the dataset with more predicted values or combine different datasets (Figure 10). This is particularly promising as efficiency is no longer the primary factor restricting its application. Research now also focuses on multi-functional integration and application of OSCs mentioned in introduction. For example, it is easier to achieve a highly transparent or highly efficient device, but it is more difficult to achieve these two characteristics simultaneously as well as stability, processability, and mechanical properties. However, this dataset and algorithm will act as a powerful tool to guide the design of multi-functional devices. ML finds rules from the beginning to the end. It will be relatively accurate with more data and important features. It would be difficult for human beings to learn and consider all kinds of processing conditions, properties, and performances at the same time using the ab initio method, especially when the materials are in the external complex environment.

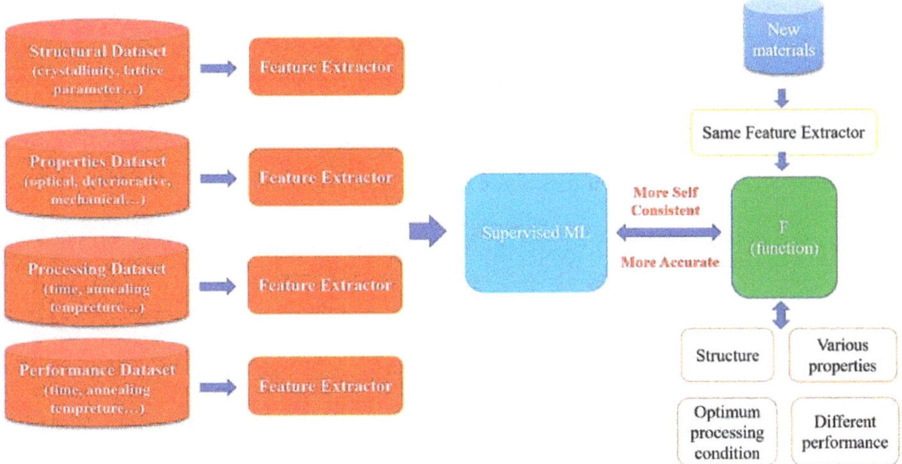

Figure 10. Basic strategies for combining structures, properties, processing, and performance via machine learning.

(c) Aid Theories Behind

Understanding the theory especially in atomic and even electronic levels behind the devices plays a significant role. It would involve the quantum mechanism and simulation of the Schrodinger equation. Although we can apply the first principle and some software packages, it is very challenging and time-consuming to build from the bottom to the top. However, we can use machine learning to guide it. For example, we can know the weights end even hidden equations between possible variables and the value of FF. These weights and equations can be an important guide for the principle establishment and material design especially when we do not fully understand the factors of a phenomenon.

7. Conclusions

Recent years have witnessed remarkable efficiency improvement in organic solar cells since the high efficiency of Y6 NFA was reported in 2019. This review started with the necessary working mechanism and summarized the key design strategies, structures, and various properties on the PM6:Y6-based OSCs for the past three years. Modifying the Y6 material by tuning the side-chain, increasing or decreasing the number of rings of the fused unit, altering the end unit, and changing atoms could serve as effective methods to further improve the PCE. As more and more data are available, and with the rapid improvement in computer calculation speed and ML algorithms, the solutions offered by ML may shed some light for the future development of highly efficient and multi-functional OSCs or other applicable fields. By improving the technologies and overcoming the challenges such as low-quality data and less-efficient feature extractors, an era of more machine filtering and aiding before experiments would be achieved in the near future.

Author Contributions: Conceptualization, S.Z.; writing—review and editing, S.Z.; supervision, B.K.Y., Z.Z. and L.Y. All authors have read and agreed to the published version of the manuscript.

Funding: This work was financially supported by the International School of Advanced Materials, South China University of Technology, Program for Science and Technology Development of Dongguan (2019622163009), the Dongguan Innovative Research Team Program (No. 2018607201002), and the Guangdong Provincial Training Program of Innovation and Entrepreneurship for Undergraduates (grant No. S202010561186). B.K.Y. thanks the financial support from Uniten bold grant J515005002/20021170. Z. Zhong thanks the financial support from the Natural Science Foundation of

China (No. 22005102), the China Postdoctoral Science Foundation (2020M682696), and the Science and Technology Program of Guangdong (No. 2020A1515011028).

Institutional Review Board Statement: Not applicable.

Informed Consent Statement: Not applicable.

Data Availability Statement: Not applicable.

Conflicts of Interest: The authors declare no conflict of interest.

References

1. Gueymard, C.A. The sun's total and spectral irradiance for solar energy applications and solar radiation models. *Sol. Energy* **2004**, *76*, 423–453. [CrossRef]
2. Britt, J.; Ferekides, C. Thin-film CdS/CdTe solar cell with 15.8% efficiency. *Appl. Phys. Lett.* **1993**, *62*, 2851–2852. [CrossRef]
3. Goetzberger, A.; Hebling, C.; Schock, H. Photovoltaic materials, history, status and outlook. *Mater. Sci. Eng. R Rep.* **2003**, *40*, 1–46. [CrossRef]
4. Sun, C.; Xia, R.; Shi, H.; Yao, H.; Liu, X.; Hou, J.; Huang, F.; Yip, H.; Cao, Y. Heat-Insulating Multifunctional Semitransparent Polymer Solar Cells. *Joule* **2018**, *2*, 1816–1826. [CrossRef]
5. Shi, H.; Xia, R.; Zhang, G.; Yip, H.; Cao, Y. Spectral Engineering of Semitransparent Polymer Solar Cells for Greenhouse Applications. *Adv. Energy Mater.* **2019**, *9*, 1803438. [CrossRef]
6. Han, Y.W.; Jeon, S.J.; Lee, H.S.; Park, H.; Kim, K.S.; Lee, H.W.; Moon, D.K. Evaporation-Free Nonfullerene Flexible Organic Solar Cell Modules Manufactured by An All-Solution Process. *Adv. Energy Mater.* **2019**, *9*, 1902065. [CrossRef]
7. Dong, S.; Jia, T.; Zhang, K.; Jing, J.; Huang, F. Single-Component Non-halogen Solvent-Processed High-Performance Organic Solar Cell Module with Efficiency over 14%. *Joule* **2020**, *4*, 2004–2016. [CrossRef]
8. Park, S.H.; Roy, A.; Beaupré, S.; Cho, S.; Coates, N.; Ji, S.M.; Moses, D.; Leclerc, M.; Lee, K.; Heeger, A.J. Bulk heterojunction solar cells with internal quantum efficiency approaching 100%. *Nat. Photonics* **2009**, *3*, 297–302. [CrossRef]
9. Sariciftci, N.S.; Smilowitz, L.; Heeger, A.J.; Wudl, F. Photoinduced electron transfer from a conducting polymer to buckminsterfullerene. *Science* **1992**, *258*, 1474–1476. [CrossRef]
10. Blom, P.W.M.; Mihailetchi, V.D.; Koster, L.J.A.; Markov, D.E. Device Physics of Polymer:Fullerene Bulk Heterojunction Solar Cells. *Adv. Mater.* **2007**, *19*, 1551–1566. [CrossRef]
11. Dennler, G.; Scharber, M.C.; Brabec, C.J. Polymer-Fullerene Bulk-Heterojunction Solar Cells. *Adv. Mater.* **2009**, *21*, 1323–1338. [CrossRef]
12. Reese, M.O.; Nardes, A.M.; Rupert, B.L.; Larsen, R.E.; Olson, D.C.; Lloyd, M.T.; Shaheen, S.E.; Ginley, D.S.; Rumbles, G.; Kopidakis, N. Photoinduced Degradation of Polymer and Polymer-Fullerene Active Layers: Experiment and Theory. *Adv. Funct. Mater.* **2010**, *20*, 3476–3483. [CrossRef]
13. You, J.; Dou, L.; Yoshimura, K.; Kato, T.; Ohya, K.; Moriarty, T.; Emery, K.; Chen, C.; Gao, J.; Li, G.; et al. A polymer tandem solar cell with 10.6% power conversion efficiency. *Nat. Commun.* **2013**, *4*, 1446. [CrossRef] [PubMed]
14. Liu, T.; Troisi, A. What Makes Fullerene Acceptors Special as Electron Acceptors in Organic Solar Cells and How to Replace Them. *Adv. Mater.* **2013**, *25*, 1038–1041. [CrossRef]
15. Zhao, J.; Li, Y.; Yang, G.; Jiang, K.; Lin, H.; Ade, H.; Ma, W.; Yan, H. Efficient organic solar cells processed from hydrocarbon solvents. *Nat. Energy* **2016**, *1*, 15027. [CrossRef]
16. Zhan, X.; Tan, Z.; Domercq, B.; An, Z.; Zhang, X.; Barlow, S.; Li, Y.; Zhu, D.; Kippelen, B.; Marder, S.R. A High-Mobility Electron-Transport Polymer with Broad Absorption and Its Use in Field-Effect Transistors and All-Polymer Solar Cells. *J. Am. Chem. Soc.* **2007**, *129*, 7246–7247. [CrossRef]
17. Cheng, P.; Li, G.; Zhan, X.; Yang, Y. Next-generation organic photovoltaics based on non-fullerene acceptors. *Nat. Photonics* **2018**, *12*, 131–142. [CrossRef]
18. Jørgensen, M.; Norrman, K.; Gevorgyan, S.A.; Tromholt, T.; Andreasen, B.; Krebs, F.C. Stability of Polymer Solar Cells. *Adv. Mater.* **2012**, *24*, 580–612. [CrossRef]
19. Lin, Y.; Wang, J.; Zhang, Z.; Bai, H.; Li, Y.; Zhu, D.; Zhan, X. An Electron Acceptor Challenging Fullerenes for Efficient Polymer Solar Cells. *Adv. Mater.* **2015**, *27*, 1170–1174. [CrossRef] [PubMed]
20. Yuan, J.; Zhang, Y.; Zhou, L.; Zhang, G.; Yip, H.; Lau, T.; Lu, X.; Zhu, C.; Peng, H.; Johnson, P.A.; et al. Single-Junction Organic Solar Cell with over 15% Efficiency Using Fused-Ring Acceptor with Electron-Deficient Core. *Joule* **2019**, *3*, 1140–1151. [CrossRef]
21. Cui, Y.; Yao, H.; Zhang, J.; Zhang, T.; Wang, Y.; Hong, L.; Xian, K.; Xu, B.; Zhang, S.; Peng, J.; et al. Over 16% efficiency organic photovoltaic cells enabled by a chlorinated acceptor with increased open-circuit voltages. *Nat. Commun.* **2019**, *10*, 2515. [CrossRef] [PubMed]
22. Li, C.; Zhou, J.; Song, J.; Xu, J.; Zhang, H.; Zhang, X.; Guo, J.; Zhu, L.; Wei, D.; Han, G.; et al. Non-fullerene acceptors with branched side chains and improved molecular packing to exceed 18% efficiency in organic solar cells. *Nat. Energy* **2021**, *6*, 605–613. [CrossRef]

23. Zhang, G.; Chen, X.; Xiao, J.; Chow, P.C.Y.; Ren, M.; Kupgan, G.; Jiao, X.; Chan, C.C.S.; Du, X.; Xia, R.; et al. Delocalization of exciton and electron wavefunction in non-fullerene acceptor molecules enables efficient organic solar cells. *Nat. Commun.* **2020**, *11*, 3943. [CrossRef]
24. Li, Y.; He, C.; Zuo, L.; Zhao, F.; Zhan, L.; Li, X.; Xia, R.; Yip, H.L.; Li, C.Z.; Liu, X.; et al. High-Performance Semi-Transparent Organic Photovoltaic Devices via Improving Absorbing Selectivity. *Adv. Energy Mater.* **2021**, *11*, 2003408. [CrossRef]
25. Xu, C.; Jin, K.; Xiao, Z.; Zhao, Z.; Ma, X.; Wang, X.; Li, J.; Xu, W.; Zhang, S.; Ding, L.; et al. Wide Bandgap Polymer with Narrow Photon Harvesting in Visible Light Range Enables Efficient Semitransparent Organic Photovoltaics. *Adv. Funct. Mater.* **2021**, *31*, 2107934. [CrossRef]
26. Lee, J.; Jeong, D.; Kim, D.J.; Phan, T.N.; Park, J.S.; Kim, T.; Kim, B.J. Flexible-spacer incorporated polymer donors enable superior blend miscibility for high-performance and mechanically-robust polymer solar cells. *Energy Environ. Sci.* **2021**, *14*, 4067–4076. [CrossRef]
27. Bernardo, G.; Lopes, T.; Lidzey, D.G.; Mendes, A. Progress in Upscaling Organic Photovoltaic Devices. *Adv. Energy Mater.* **2021**, *11*, 2100342. [CrossRef]
28. Brabec, C.J.; Distler, A.; Du, X.; Egelhaaf, H.J.; Hauch, J.; Heumueller, T.; Li, N. Material Strategies to Accelerate OPV Technology Toward a GW Technology. *Adv. Energy Mater.* **2020**, *10*, 2001864. [CrossRef]
29. Xu, Y.; Yao, H.; Ma, L.; Wu, Z.; Cui, Y.; Hong, L.; Zu, Y.; Wang, J.; Woo, H.Y.; Hou, J. Organic photovoltaic cells with high efficiencies for both indoor and outdoor applications. *Mater. Chem. Front.* **2021**, *5*, 893–900. [CrossRef]
30. Yin, Z.; Wei, J.; Zheng, Q. Interfacial Materials for Organic Solar Cells: Recent Advances and Perspectives. *Adv. Sci.* **2016**, *3*, 1500362. [CrossRef]
31. Mishra, A. Interfacial Materials for Organic Solar Cells. In *Solar Energy: Systems, Challenges, and Opportunities*; Tyagi, H., Chakraborty, P.R., Powar, S., Agarwal, A.K., Eds.; Springer: Singapore, 2020; pp. 373–423.
32. Kumaresan, P.; Vegiraju, S.; Ezhumalai, Y.; Yau, S.; Kim, C.; Lee, W.; Chen, M. Fused-Thiophene Based Materials for Organic Photovoltaics and Dye-Sensitized Solar Cells. *Polymers* **2014**, *6*, 2645–2669. [CrossRef]
33. Heeger, A.J. 25th Anniversary Article: Bulk Heterojunction Solar Cells: Understanding the Mechanism of Operation. *Adv. Mater.* **2014**, *26*, 10–28. [CrossRef]
34. Gupta, D.; Mukhopadhyay, S.; Narayan, K.S. Fill factor in organic solar cells. *Sol. Energy Mater. Sol. Cells* **2010**, *94*, 1309–1313. [CrossRef]
35. Nagasawa, S.; Al-Naamani, E.; Saeki, A. Computer-Aided Screening of Conjugated Polymers for Organic Solar Cell: Classification by Random Forest. *J. Phys. Chem. Lett.* **2018**, *9*, 2639–2646. [CrossRef]
36. Li, N.; McCulloch, I.; Brabec, C.J. Analyzing the efficiency, stability and cost potential for fullerene-free organic photovoltaics in one figure of merit. *Energy Environ. Sci.* **2018**, *11*, 1355–1361. [CrossRef]
37. Yao, J.; Kirchartz, T.; Vezie, M.S.; Faist, M.A.; Gong, W.; He, Z.; Wu, H.; Troughton, J.; Watson, T.; Bryant, D.; et al. Quantifying Losses in Open-Circuit Voltage in Solution-Processable Solar Cells. *Phys. Rev. Appl.* **2015**, *4*, 14020. [CrossRef]
38. Wang, J.; Yao, H.; Xu, J.; Hou, J. Recent Progress in Reducing Voltage Loss in Organic Photovoltaic Cells. *Mater. Chem. Front.* **2021**, *5*, 709–722. [CrossRef]
39. Hong, L.; Yao, H.; Wu, Z.; Cui, Y.; Zhang, T.; Xu, Y.; Yu, R.; Liao, Q.; Gao, B.; Xian, K.; et al. Eco-Compatible Solvent-Processed Organic Photovoltaic Cells with Over 16% Efficiency. *Adv. Mater.* **2019**, *31*, 1903441. [CrossRef]
40. Fan, B.; Zhang, D.; Li, M.; Zhong, W.; Zeng, Z.; Ying, L.; Huang, F.; Cao, Y. Achieving over 16% efficiency for single-junction organic solar cells. *Sci. China Chem.* **2019**, *62*, 746–752. [CrossRef]
41. Li, W.; Ye, L.; Li, S.; Yao, H.; Ade, H.; Hou, J. A High-Efficiency Organic Solar Cell Enabled by the Strong Intramolecular Electron Push–Pull Effect of the Nonfullerene Acceptor. *Adv. Mater.* **2018**, *30*, 1707170. [CrossRef]
42. Yuan, J.; Zhang, Y.; Zhou, L.; Zhang, C.; Lau, T.K.; Zhang, G.; Lu, X.; Yip, H.L.; So, S.K.; Beaupré, S.; et al. Fused Benzothiadiazole: A Building Block for n-Type Organic Acceptor to Achieve High-Performance Organic Solar Cells. *Adv. Mater.* **2019**, *31*, 1807577. [CrossRef] [PubMed]
43. Yuan, J.; Huang, T.; Cheng, P.; Zou, Y.; Zhang, H.; Yang, J.L.; Chang, S.; Zhang, Z.; Huang, W.; Wang, R.; et al. Enabling low voltage losses and high photocurrent in fullerene-free organic photovoltaics. *Nat. Commun.* **2019**, *10*, 570. [CrossRef] [PubMed]
44. Cheng, P.; Yang, Y. Narrowing the Band Gap: The Key to High-Performance Organic Photovoltaics. *Acc. Chem. Res.* **2020**, *53*, 1218–1228. [CrossRef]
45. Feng, L.; Yuan, J.; Zhang, Z.; Peng, H.; Zhang, Z.; Xu, S.; Liu, Y.; Li, Y.; Zou, Y. Thieno[3,2-b]pyrrolo-Fused Pentacyclic Benzotriazole-Based Acceptor for Efficient Organic Photovoltaics. *ACS Appl. Mater. Interfaces* **2017**, *9*, 31985–31992. [CrossRef] [PubMed]
46. Cheng, Y.; Chen, C.; Ho, Y.; Chang, S.; Witek, H.A.; Hsu, C. Thieno[3,2-b]pyrrolo Donor Fused with Benzothiadiazolo, Benzoselenadiazolo and Quinoxalino Acceptors: Synthesis, Characterization, and Molecular Properties. *Org. Lett.* **2011**, *13*, 5484–5487. [CrossRef]
47. Zhang, D.; Fan, B.; Ying, L.; Li, N.; Brabec, C.J.; Huang, F.; Cao, Y. Recent progress in thick-film organic photovoltaic devices: Materials, devices, and processing. *SusMat* **2021**, *1*, 4–23. [CrossRef]
48. Chen, Z.; Cai, P.; Chen, J.; Liu, X.; Zhang, L.; Lan, L.; Peng, J.; Ma, Y.; Cao, Y. Low Band-Gap Conjugated Polymers with Strong Interchain Aggregation and Very High Hole Mobility Towards Highly Efficient Thick-Film Polymer Solar Cells. *Adv. Mater.* **2014**, *26*, 2586–2591. [CrossRef]

49. Lei, T.; Xia, X.; Wang, J.; Liu, C.; Pei, J. "Conformation Locked" Strong Electron-Deficient Poly(p-Phenylene Vinylene) Derivatives for Ambient-Stable n-Type Field-Effect Transistors: Synthesis, Properties, and Effects of Fluorine Substitution Position. *J. Am. Chem. Soc.* **2014**, *136*, 2135–2141. [CrossRef]
50. Huang, H.; Yang, L.; Facchetti, A.; Marks, T.J. Organic and Polymeric Semiconductors Enhanced by Noncovalent Conformational Locks. *Chem. Rev.* **2017**, *117*, 10291–10318. [CrossRef]
51. Zheng, Z.; Yao, H.; Ye, L.; Xu, Y.; Zhang, S.; Hou, J. PBDB-T and its derivatives: A family of polymer donors enables over 17% efficiency in organic photovoltaics. *Mater. Today* **2020**, *35*, 115–130. [CrossRef]
52. Wang, R.; Yuan, J.; Wang, R.; Han, G.; Huang, T.; Huang, W.; Xue, J.; Wang, H.C.; Zhang, C.; Zhu, C.; et al. Rational Tuning of Molecular Interaction and Energy Level Alignment Enables High-Performance Organic Photovoltaics. *Adv. Mater.* **2019**, *31*, 1904215. [CrossRef] [PubMed]
53. Wang, J.; Zhan, X. Fused-Ring Electron Acceptors for Photovoltaics and Beyond. *Acc. Chem. Res.* **2021**, *54*, 132–143. [CrossRef] [PubMed]
54. Ma, X.; Tang, C.; Ma, Y.; Zhu, X.; Wang, J.; Gao, J.; Xu, C.; Wang, Y.; Zhang, J.; Zheng, Q.; et al. Over 17% Efficiency of Ternary Organic Photovoltaics Employing Two Acceptors with an Acceptor–Donor–Acceptor Configuration. *ACS Appl. Mater. Interfaces* **2021**, *13*, 57684–57692. [CrossRef] [PubMed]
55. He, Z.; Zhong, C.; Su, S.; Xu, M.; Wu, H.; Cao, Y. Enhanced power-conversion efficiency in polymer solar cells using an inverted device structure. *Nat. Photonics* **2012**, *6*, 591–595. [CrossRef]
56. Wang, Y.; Bai, H.; Zhan, X. Comparison of conventional and inverted structures in fullerene-free organic solar cells. *J. Energy Chem.* **2015**, *24*, 744–749. [CrossRef]
57. Fan, B.; Li, M.; Zhang, D.; Zhong, W.; Ying, L.; Zeng, Z.; An, K.; Huang, Z.; Shi, L.; Bazan, G.C.; et al. Tailoring Regioisomeric Structures of π-Conjugated Polymers Containing Monofluorinated π-Bridges for Highly Efficient Polymer Solar Cells. *ACS Energy Lett.* **2020**, *5*, 2087–2094. [CrossRef]
58. Ma, R.; Liu, T.; Luo, Z.; Guo, Q.; Xiao, Y.; Chen, Y.; Li, X.; Luo, S.; Lu, X.; Zhang, M.; et al. Improving open-circuit voltage by a chlorinated polymer donor endows binary organic solar cells efficiencies over 17%. *Sci. China Chem.* **2020**, *63*, 325–330. [CrossRef]
59. Liu, Q.; Jiang, Y.; Jin, K.; Qin, J.; Xu, J.; Li, W.; Xiong, J.; Liu, J.; Xiao, Z.; Sun, K.; et al. 18% Efficiency organic solar cells. *Sci. Bull.* **2020**, *65*, 272–275. [CrossRef]
60. Yan, T.; Song, W.; Huang, J.; Peng, R.; Huang, L.; Ge, Z. 16.67% Rigid and 14.06% Flexible Organic Solar Cells Enabled by Ternary Heterojunction Strategy. *Adv. Mater.* **2019**, *31*, 1902210. [CrossRef]
61. Yu, R.; Yao, H.; Cui, Y.; Hong, L.; He, C.; Hou, J. Improved Charge Transport and Reduced Nonradiative Energy Loss Enable Over 16% Efficiency in Ternary Polymer Solar Cells. *Adv. Mater.* **2019**, *31*, 1902302. [CrossRef]
62. Li, K.; Wu, Y.; Tang, Y.; Pan, M.A.; Ma, W.; Fu, H.; Zhan, C.; Yao, J. Ternary Blended Fullerene-Free Polymer Solar Cells with 16.5% Efficiency Enabled with a Higher-LUMO-Level Acceptor to Improve Film Morphology. *Adv. Energy Mater.* **2019**, *9*, 1901728. [CrossRef]
63. Wang, X.; Sun, Q.; Gao, J.; Wang, J.; Xu, C.; Ma, X.; Zhang, F. Recent Progress of Organic Photovoltaics with Efficiency over 17%. *Energies* **2021**, *14*, 4200. [CrossRef]
64. Xu, W.; Ma, X.; Son, J.H.; Jeong, S.Y.; Niu, L.; Xu, C.; Zhang, S.; Zhou, Z.; Gao, J.; Woo, H.Y.; et al. Smart Ternary Strategy in Promoting the Performance of Polymer Solar Cells Based on Bulk-Heterojunction or Layer-By-Layer Structure. *Small* **2021**, 2104215. [CrossRef] [PubMed]
65. Yang, J.; Geng, Y.; Li, J.; Zhao, B.; Guo, Q.; Zhou, E. A-DA′D-A-Type Non-fullerene Acceptors Containing a Fused Heptacyclic Ring for Poly(3-hexylthiophene)-Based Polymer Solar Cells. *J. Phys. Chem. C* **2020**, *124*, 24616–24623. [CrossRef]
66. Yang, C.; Zhang, S.; Ren, J.; Gao, M.; Bi, P.; Ye, L.; Hou, J. Molecular design of a non-fullerene acceptor enables a P3HT-based organic solar cell with 9.46% efficiency. *Energy Environ. Sci.* **2020**, *13*, 2864–2869. [CrossRef]
67. Zhang, Z.; Li, Y.; Cai, G.; Zhang, Y.; Lu, X.; Lin, Y. Selenium Heterocyclic Electron Acceptor with Small Urbach Energy for As-Cast High-Performance Organic Solar Cells. *J. Am. Chem. Soc.* **2020**, *142*, 18741–18745. [CrossRef]
68. Zhu, C.; An, K.; Zhong, W.; Li, Z.; Qian, Y.; Su, X.; Ying, L. Design and synthesis of non-fullerene acceptors based on a quinoxalineimide moiety as the central building block for organic solar cells. *Chem. Commun.* **2020**, *56*, 4700–4703. [CrossRef] [PubMed]
69. Wang, G.; Adil, M.A.; Zhang, J.; Wei, Z. Large-Area Organic Solar Cells: Material Requirements, Modular Designs, and Printing Methods. *Adv. Mater.* **2018**, *31*, 1805089. [CrossRef]
70. Zhang, H.; Yao, H.; Hou, J.; Zhu, J.; Zhang, J.; Li, W.; Yu, R.; Gao, B.; Zhang, S.; Hou, J. Over 14% Efficiency in Organic Solar Cells Enabled by Chlorinated Nonfullerene Small-Molecule Acceptors. *Adv. Mater.* **2018**, *30*, 1800613. [CrossRef]
71. Zhang, Y.; Yao, H.; Zhang, S.; Qin, Y.; Zhang, J.; Yang, L.; Li, W.; Wei, Z.; Gao, F.; Hou, J. Fluorination vs. chlorination: A case study on high performance organic photovoltaic materials. *Sci. China Chem.* **2018**, *61*, 1328–1337. [CrossRef]
72. Cai, F.; Zhu, C.; Yuan, J.; Li, Z.; Meng, L.; Liu, W.; Peng, H.; Jiang, L.; Li, Y.; Zou, Y. Efficient organic solar cells based on a new "Y-series" non-fullerene acceptor with an asymmetric electron-deficient-core. *Chem. Commun.* **2020**, *56*, 4340–4343. [CrossRef] [PubMed]
73. Gao, W.; Fu, H.; Li, Y.; Lin, F.; Sun, R.; Wu, Z.; Wu, X.; Zhong, C.; Min, J.; Luo, J.; et al. Asymmetric Acceptors Enabling Organic Solar Cells to Achieve an over 17% Efficiency: Conformation Effects on Regulating Molecular Properties and Suppressing Nonradiative Energy Loss. *Adv. Energy Mater.* **2021**, *11*, 2003177. [CrossRef]

74. Gao, W.; Fan, B.; Qi, F.; Lin, F.; Sun, R.; Xia, X.; Gao, J.; Zhong, C.; Lu, X.; Min, J.; et al. Asymmetric Isomer Effects in Benzo[c][1,2,5]thiadiazole-Fused Nonacyclic Acceptors: Dielectric Constant and Molecular Crystallinity Control for Significant Photovoltaic Performance Enhancement. *Adv. Funct. Mater.* **2021**, *31*, 2104369. [CrossRef]
75. Chai, G.; Chang, Y.; Zhang, J.; Xu, X.; Yu, L.; Zou, X.; Li, X.; Chen, Y.; Luo, S.; Liu, B.; et al. Fine-tuning of side-chain orientations on nonfullerene acceptors enables organic solar cells with 17.7% efficiency. *Energy Environ. Sci.* **2021**, *14*, 3469–3479. [CrossRef]
76. Ridley, J.; Zerner, M. An intermediate neglect of differential overlap technique for spectroscopy: Pyrrole and the azines. *Theor. Chim. Acta* **1973**, *32*, 111–134. [CrossRef]
77. He, C.; Li, Y.; Liu, Y.; Li, Y.; Zhou, G.; Li, S.; Zhu, H.; Lu, X.; Zhang, F.; Li, C.; et al. Near infrared electron acceptors with a photoresponse beyond 1000 nm for highly efficient organic solar cells. *J. Mater. Chem. A* **2020**, *8*, 18154–18161. [CrossRef]
78. Hai, J.; Zhao, W.; Luo, S.; Yu, H.; Chen, H.; Lu, Z.; Li, L.; Zou, Y.; Yan, H. Vinylene π-bridge: A simple building block for ultra-narrow bandgap nonfullerene acceptors enable 14.2% efficiency in binary organic solar cells. *Dye. Pigment.* **2021**, *188*, 109171. [CrossRef]
79. Jiang, K.; Wei, Q.; Lai, J.Y.L.; Peng, Z.; Kim, H.K.; Yuan, J.; Ye, L.; Ade, H.; Zou, Y.; Yan, H. Alkyl Chain Tuning of Small Molecule Acceptors for Efficient Organic Solar Cells. *Joule* **2019**, *3*, 3020–3033. [CrossRef]
80. Cui, Y.; Yao, H.; Hong, L.; Zhang, T.; Tang, Y.; Lin, B.; Xian, K.; Gao, B.; An, C.; Bi, P.; et al. Organic photovoltaic cell with 17% efficiency and superior processability. *Natl. Sci. Rev.* **2020**, *7*, 1239–1246. [CrossRef]
81. Yu, G.; Gao, J.; Hummelen, J.C.; Wudl, F.; Heeger, A.J. Polymer Photovoltaic Cells: Enhanced Efficiencies via a Network of Internal Donor-Acceptor Heterojunctions. *Science* **1995**, *270*, 1789–1791. [CrossRef]
82. Halls, J.; Walsh, C.A.; Greenham, N.C.; Marseglia, E.A.; Friend, R.H.; Moratti, S.C.; Holmes, A.B. Efficient photodiodes from interpenetrating polymer networks. *Nature* **1995**, *376*, 498–500. [CrossRef]
83. Moore, J.R.; Albert-Seifried, S.; Rao, A.; Massip, S.; Watts, B.; Morgan, D.J.; Friend, R.H.; McNeill, C.R.; Sirringhaus, H. Polymer Blend Solar Cells Based on a High-Mobility Naphthalenediimide-Based Polymer Acceptor: Device Physics, Photophysics and Morphology. *Adv. Energy Mater.* **2011**, *1*, 230–240. [CrossRef]
84. Fan, B.; Ying, L.; Zhu, P.; Pan, F.; Liu, F.; Chen, J.; Huang, F.; Cao, Y. All-Polymer Solar Cells Based on a Conjugated Polymer Containing Siloxane-Functionalized Side Chains with Efficiency over 10%. *Adv. Mater.* **2017**, *29*, 1703906. [CrossRef] [PubMed]
85. Jia, T.; Zhang, J.; Zhong, W.; Liang, Y.; Zhang, K.; Dong, S.; Ying, L.; Liu, F.; Wang, X.; Huang, F.; et al. 14.4% efficiency all-polymer solar cell with broad absorption and low energy loss enabled by a novel polymer acceptor. *Nano Energy* **2020**, *72*, 104718. [CrossRef]
86. Wang, W.; Wu, Q.; Sun, R.; Guo, J.; Wu, Y.; Shi, M.; Yang, W.; Li, H.; Min, J. Controlling Molecular Mass of Low-Band-Gap Polymer Acceptors for High-Performance All-Polymer Solar Cells. *Joule* **2020**, *4*, 1070–1086. [CrossRef]
87. Yu, H.; Qi, Z.; Yu, J.; Xiao, Y.; Sun, R.; Luo, Z.; Cheung, A.M.H.; Zhang, J.; Sun, H.; Zhou, W.; et al. Fluorinated End Group Enables High-Performance All-Polymer Solar Cells with Near-Infrared Absorption and Enhanced Device Efficiency over 14%. *Adv. Energy Mater.* **2021**, *11*, 2003171. [CrossRef]
88. Peng, F.; An, K.; Zhong, W.; Li, Z.; Ying, L.; Li, N.; Huang, Z.; Zhu, C.; Fan, B.; Huang, F.; et al. A Universal Fluorinated Polymer Acceptor Enables All-Polymer Solar Cells with >15% Efficiency. *ACS Energy Lett.* **2020**, *5*, 3702–3707. [CrossRef]
89. Yu, H.; Pan, M.; Sun, R.; Agunawela, I.; Zhang, J.; Li, Y.; Qi, Z.; Han, H.; Zou, X.; Zhou, W.; et al. Regio-Regular Polymer Acceptors Enabled by Determined Fluorination on End Groups for All-Polymer Solar Cells with 15.2% Efficiency. *Angew. Chem.* **2021**, *133*, 10225–10234. [CrossRef]
90. Yu, H.; Luo, S.; Sun, R.; Angunawela, I.; Qi, Z.; Peng, Z.; Zhou, W.; Han, H.; Wei, R.; Pan, M.; et al. A Difluoro-Monobromo End Group Enables High-Performance Polymer Acceptor and Efficient All-Polymer Solar Cells Processable with Green Solvent under Ambient Condition. *Adv. Funct. Mater.* **2021**, *31*, 2100791. [CrossRef]
91. Sun, R.; Wang, W.; Yu, H.; Chen, Z.; Xia, X.; Shen, H.; Guo, J.; Shi, M.; Zheng, Y.; Wu, Y.; et al. Achieving over 17% efficiency of ternary all-polymer solar cells with two well-compatible polymer acceptors. *Joule* **2021**, *5*, 1548–1565. [CrossRef]
92. Butler, K.T.; Davies, D.W.; Cartwright, H.; Isayev, O.; Walsh, A. Machine learning for molecular and materials science. *Nature* **2018**, *559*, 547–555. [CrossRef] [PubMed]
93. Ghosh, A.K.; Feng, T. Rectification, space-charge-limited current, photovoltaic and photoconductive properties of Al/tetracene/Au sandwich cell. *J. Appl. Phys.* **1973**, *44*, 2781–2788. [CrossRef]
94. Sun, W.; Zheng, Y.; Yang, K.; Zhang, Q.; Shah, A.A.; Wu, Z.; Sun, Y.; Feng, L.; Chen, D.; Xiao, Z.; et al. Machine learning-assisted molecular design and efficiency prediction for high-performance organic photovoltaic materials. *Sci. Adv.* **2019**, *5*, 4275. [CrossRef] [PubMed]
95. Hachmann, J.; Olivares-Amaya, R.; Atahan-Evrenk, S.; Amador-Bedolla, C.; Sánchez-Carrera, R.S.; Gold-Parker, A.; Vogt, L.; Brockway, A.M.; Aspuru-Guzik, A. The Harvard Clean Energy Project: Large-Scale Computational Screening and Design of Organic Photovoltaics on the World Community Grid. *J. Phys. Chem. Lett.* **2011**, *2*, 2241–2251. [CrossRef]
96. Pyzer-Knapp, E.O.; Li, K.; Aspuru-Guzik, A. Learning from the Harvard Clean Energy Project: The Use of Neural Networks to Accelerate Materials Discovery. *Adv. Funct. Mater.* **2015**, *25*, 6495–6502. [CrossRef]
97. Lopez, S.A.; Pyzer-Knapp, E.O.; Simm, G.N.; Lutzow, T.; Li, K.; Seress, L.R.; Hachmann, J.; Aspuru-Guzik, A. The Harvard organic photovoltaic dataset. *Sci. Data* **2016**, *3*, 160086. [CrossRef]
98. Sahu, H.; Rao, W.; Troisi, A.; Ma, H. Toward Predicting Efficiency of Organic Solar Cells via Machine Learning and Improved Descriptors. *Adv. Energy Mater.* **2018**, *8*, 1801032. [CrossRef]

99. Jørgensen, P.B.; Mesta, M.; Shil, S.; García Lastra, J.M.; Jacobsen, K.W.; Thygesen, K.S.; Schmidt, M.N. Machine learning-based screening of complex molecules for polymer solar cells. *J. Chem. Phys.* **2018**, *148*, 241735. [CrossRef]
100. Padula, D.; Simpson, J.D.; Troisi, A. Combining electronic and structural features in machine learning models to predict organic solar cells properties. *Mater. Horiz.* **2019**, *6*, 343–349. [CrossRef]
101. Sun, W.; Li, M.; Li, Y.; Wu, Z.; Sun, Y.; Lu, S.; Xiao, Z.; Zhao, B.; Sun, K. The Use of Deep Learning to Fast Evaluate Organic Photovoltaic Materials. *Adv. Theory Simul.* **2019**, *2*, 1800116. [CrossRef]
102. Du, X.; Lüer, L.; Heumueller, T.; Wagner, J.; Berger, C.; Osterrieder, T.; Wortmann, J.; Langner, S.; Vongsaysy, U.; Bertrand, M.; et al. Elucidating the Full Potential of OPV Materials Utilizing a High-Throughput Robot-Based Platform and Machine Learning. *Joule* **2021**, *5*, 495–506. [CrossRef]
103. Cambria, E.; White, B. Jumping NLP Curves: A Review of Natural Language Processing Research [Review Article]. *IEEE Comput. Intell. Mag.* **2014**, *9*, 48–57. [CrossRef]
104. Cereto-Massagué, A.; Ojeda, M.J.; Valls, C.; Mulero, M.; Garcia-Vallvé, S.; Pujadas, G. Molecular fingerprint similarity search in virtual screening. *Methods* **2015**, *71*, 58–63. [CrossRef] [PubMed]
105. Durant, J.L.; Leland, B.A.; Henry, D.R.; Nourse, J.G. Reoptimization of MDL Keys for Use in Drug Discovery. *J. Chem. Inf. Comput. Sci.* **2002**, *42*, 1273–1280. [CrossRef]
106. Tovar, A.; Eckert, H.; Bajorath, J. Comparison of 2D Fingerprint Methods for Multiple-Template Similarity Searching on Compound Activity Classes of Increasing Structural Diversity. *ChemMedChem* **2007**, *2*, 208–217. [CrossRef]
107. Rester, U. From virtuality to reality—Virtual screening in lead discovery and lead optimization: A medicinal chemistry perspective. *Curr. Opin. Drug Discov. Dev.* **2008**, *11*, 559.
108. Liang, L.; Liu, M.; Martin, C.; Elefteriades, J.A.; Sun, W. A machine learning approach to investigate the relationship between shape features and numerically predicted risk of ascending aortic aneurysm. *Biomech. Model. Mechanobiol.* **2017**, *16*, 1519–1533. [CrossRef]
109. Keogh, E.; Mueen, A. Curse of Dimensionality. *Ind. Eng. Chem.* **2009**, *29*, 48–53.

MDPI

Article

Substrate Effects on the Random Lasing Performance of Solution-Processed Hybrid-Perovskite Multicrystal Film

Jingyun Hu and Xinping Zhang *

Institute of Information Photonics Technology, Beijing University of Technology, Beijing 100124, China; hujingyun@emails.bjut.edu.cn
* Correspondence: zhangxinping@bjut.edu.cn

Abstract: We report dependence of random lasing performance of directly spin-coated multicrystalline thin films of an organic–inorganic hybrid, halide perovskite $CH_3NH_3PbBr_3$ ($MAPbBr_3$), on different substrates. It was discovered that random lasing performance is strongly dependent on the surface energy properties of the substrate, which determine the morphology and crystallization properties of the spin-coated film, and will consequently determine its optical scattering and emission properties. Using indium–tin oxide (ITO)-coated glass, fused silica, and tricyclo[5.2.1.$0^{2,6}$] decanedimethanol diacrylate (ADCP)-coated fused silica as the substrate materials, we compared the spectroscopic response of the random lasers and thus justified the photophysical mechanisms involved. The modification of the surface properties of the substrate enables controlling of the $MAPbBr_3$ crystallization and leads to the changing of the random lasing properties. The discoveries herein are also important for the construction of other types of laser devices, where the substrate effects should be considered during the design and preparation of the micro-/nano structures.

Keywords: substrate effects; random lasers; surface morphology; multicrystalline film of $MAPbBr_3$; contact angles

Citation: Hu, J.; Zhang, X. Substrate Effects on the Random Lasing Performance of Solution-Processed Hybrid-Perovskite Multicrystal Film. *Crystals* **2022**, *12*, 334. https://doi.org/10.3390/cryst12030334

Academic Editor: Dmitry Medvedev

Received: 1 February 2022
Accepted: 23 February 2022
Published: 27 February 2022

1. Introduction

Lasing performance and devices in various forms have been explored extensively for hybrid organic–inorganic perovskites (HOIPs) [1–6]. Random lasers based on these materials are also reported on the basis of different random structures and optical mechanisms [7–10]. Multicrystalline structures with randomly distributed interfaces have supplied the optical scattering centers in most of the cases. Although random lasers are categorized into coherent [11–14] and incoherent types [15–17], reported random lasing processes using HOIPs are mostly based on incoherent processes, so that broad-band emission spectra are observed for these lasers.

The green-emitting $CH_3NH_3PbBr_3$ ($MAPbBr_3$) is one of the most popular HOIP materials for investigations on lasing devices [18–20] because of its high emission efficiency, excellent stability, and easy synthesizing and processing advantages. $MAPbBr_3$ is a directbandgap material with a bandgap energy of about 2.3 eV, which may be excited to produce green light emission. Defects in the crystalline structures of this material may induce slight tuning of the energy band structures [21,22]. However, related studies have mainly focused on the design of the materials, their microstructures, fabrication techniques, and photophysical properties [23–26]; substrate-related effects have not been considered so far. In this work, we investigate how the surface properties of substrates made of different materials influence the morphological and microscopic compositions of the multicrystalline film of $MAPbBr_3$, and consequently influence its spectroscopic response and random lasing performance. The revealed mechanisms will be helpful for the design and realization of optoelectronic devices, including different forms of lasers, using $MAPbBr_3$ as the active medium.

2. Thin Films of Multicrystalline $MAPbBr_3$ on Different Substrates

Multicrystalline films were produced by spin-coating a solution of $MAPbBr_3$ in N,N-dimethylformamide (DMF) with a concentration 560 mg/mL onto different substrates at a speed of 3000 rpm and a duration of 30 s. The materials for synthesizing $MAPbBr_3$ were purchased from Xi'an Polymer Light Technology Corp. (Xi'an, China) and three different substrates were used: (1) Indium–tin oxide (ITO)-coated glass substrate; (2) Fused silica (FS) substrate; (3) ADCP-coated fused silica (FS) substrate. The precursor solution of ADCP was prepared by mixing 0.7-g tricyclo[5.2.1.$0^{2,6}$] decanedimethanol diacrylate (A-DCP), 0.3-g dipentaerythritol penta-/hexa-acrylate, and 5-mg 2,2–Dimethoxy-2-phenylacetophenone (DMPA) in a 3-mL bottle, which was stirred for 2 h under red light. The solution was then spin-coated onto a clean FS substrate at a speed of 3000 rpm for 30 s and the produced ADCP film was cured under a UV lamp for about 1 min for the ADCP coating finishing process.

The reason we chose ADCP to modify the surface properties of the substrate is that the ADCP layer has no physical or chemical reactions with the DMF solutions. This can be confirmed by our test experimental results in Figure S1, where the optical microscopic images of an ADCP grating with a period of 4 μm before and after it was immersed in DMF were measured. No changes can be observed in the grating structure due to its interaction with DMF. Therefore, the ADCP-coating on the FS substrate is not destroyed and there is no contamination of $MAPbBr_3$ by ADCP. All these substrates have a thickness of about 1 mm and an area of 15×15 mm^2. We define these three types of substrates as the ITO substrate, the FS substrate, and the ADCP substrate, respectively, for the convenience of discussion in the following sections. Figure 1a–c shows the characterization of the surface properties of the three substrates and the microscopic properties of the spin-coated $MAPbBr_3$ film on them. Figure 1(a_1,b_1,c_1) shows the contact angle measurements for the $MAPbBr_3$/DMF solution on the ADCP, ITO, and FS substrates, respectively. The corresponding contact angles were measured to be $\theta_{ADCP} = 27^\circ$, $\theta_{ITO} = 12^\circ$, $\theta_{FS} = 14^\circ$, implying the strongest wetting of the solution was on the ITO substrate and the weakest was on the ADCP. However, the difference is very small between ITO and FS. Such wetting properties directly influence the crystallization properties of $MAPbBr_3$ as it is spin-coated onto these substrates. Figure 1(a_2,b_2,c_2) shows scanning electron microscope images measured on the samples with ADCP, ITO, and FS substrates, respectively, and Figure 1(a_3,b_3,c_3) shows the enlarged views. The insets in Figure 1(a_2,b_2,c_2) show a local-area view, revealing the crystallization details of $MAPbBr_3$ on these three substrates. Furthermore, pinholes and cracks can be clearly observed in Figure 1(a_2,a_3). The corresponding mechanisms for the formation of such structures can be understood by recent investigations in [22].

According to Figure 1, there are two types of microscopic structures in the multicrystalline films: (1) a randomly distributed crisscross network as a large-scale modulation and as the main feature of the surface morphology; (2) randomly distributed nanoscale crystal particles, as shown in the insets and highlighted by the yellow circles. Clearly, the FS-substrate sample exhibits the thickest wrinkle-like stripes, whereas the ADCP-substrate produces the finest surface morphology with the thinnest modulation. In particular, the modulation exhibits shallow furrows for the ADCP substrate, instead of protruding stripes as for the other substrates. These different features are highlighted by tilted rectangles in red in Figure 1(a_3,b_3,c_3). However, the largest crystal particles are produced in the ADCP-substrate samples and smallest in the FS substrate, as highlighted by yellow circles in the insets in Figure 1(a_2,b_2,c_2). It is understandable that the larger crystal particles in the ADCP-substrate-based sample imply stronger optical scattering, which favors a more efficient random lasing process.

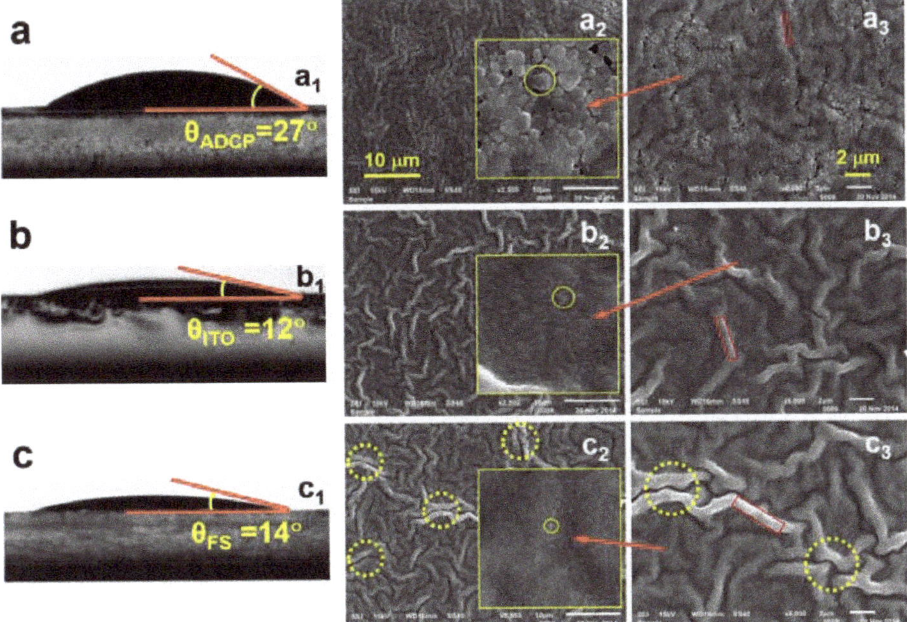

Figure 1. Substrate surface properties measured on (**a**) ADCP-coated fused silica, (**b**) ITO, and (**c**) fused silica. (**a_1**,**b_1**,**c_1**) Contact angle measurements of the DFM solution of $MAPbBr_3$ on substrates of ADCP, ITO, and FS, respectively. (**a_2**,**a_3**) SEM images of the $MAPbBr_3$-coated ADCP with different magnifications, inset: an enlarged local area. (**b_2**,**b_3**) SEM images of the $MAPbBr_3$-coated ITO with different magnifications, inset: an enlarged local area. (**c_2**,**c_3**) SEM images of the $MAPbBr_3$-coated FS with different magnifications, inset: an enlarged local area.

3. Absorption and Photoluminescence (PL) Spectroscopic Properties

Figure 2 shows the absorption and emission spectroscopic properties of the $MAPbBr_3$ multicrystalline films on different substrates. As shown in Figure 2a, the absorption spectrum has the largest amplitude for the ITO substrate and the smallest for the ADCP, if evaluating the measurement data at the peak wavelength of about 525 nm. However, the relationship between them changes when these spectra are normalized at 525 nm, as shown in the inset in Figure 2a. Since the lasing actions take place at wavelengths longer than 540 nm, we also focus our attention on this spectral range. It should be noted that the light-scattering process makes a large contribution to the absorption spectrum in this range. Therefore, the contrast of the spectral feature at 525 nm also reflects, to some extent, the strength of the scattering process. According to the inset of Figure 2a, the sample based on the ADCP substrate exhibits the strongest light-scattering properties, as can be justified by the absorption strength on both sides of the peak wavelength of 525 nm. Meanwhile, the sample based on the FS substrate exhibits the weakest optical scattering, and that on the ITO substrate is located between the other two. This also agrees well with the microscopic observations in the insets in Figure 1(a_2,b_2,c_2). As will be demonstrated in Section 4, these optical scattering processes also agree well with the random lasing properties.

Figure 2b shows the normalized PL spectra measured on the three samples with different substrates. Clearly, the PL spectrum for the ADCP substrate ($I_{ADCP}(\lambda)$) exhibits the broadest bandwidth and that for the ITO substrate ($I_{ITO}(\lambda)$) exhibits the narrowest; meanwhile, that for the FS substrate ($I_{FS}(\lambda)$) is located within the spectral range of shortest wavelengths. In the Figure 2b inset, we show the calculation results of $I_{ADCP}(\lambda)/I_{ITO}(\lambda)$ and $I_{ADCP}(\lambda)/I_{FS}(\lambda)$ by the black and red empty circles, respectively. There is a relative enhancement factor of about 1.25 in the PL spectrum by the ADCP-substrate sample at about 556.5 nm with respect to the ITO- and FS-substrate samples, as also highlighted by the arrow in magenta in Figure 2b. This is a stable and well-reproducible mechanism, implying that stronger de-wetting of the solution on the substrate and the consequently larger crystal particles are responsible for the red-shifted and broadened emission spectrum, which also explains the substrate-dependence of the random lasing performance in the subsequent sections. We must stress that the modulation on the PL spectra in Figure 2b did not result from the substrate emission. As shown in Figure S2, where we measured the PL spectra from the three substrates when they are excited by 400-nm femtosecond pulses, no emission from them can be observed. The measured spectra are only composed of background noise.

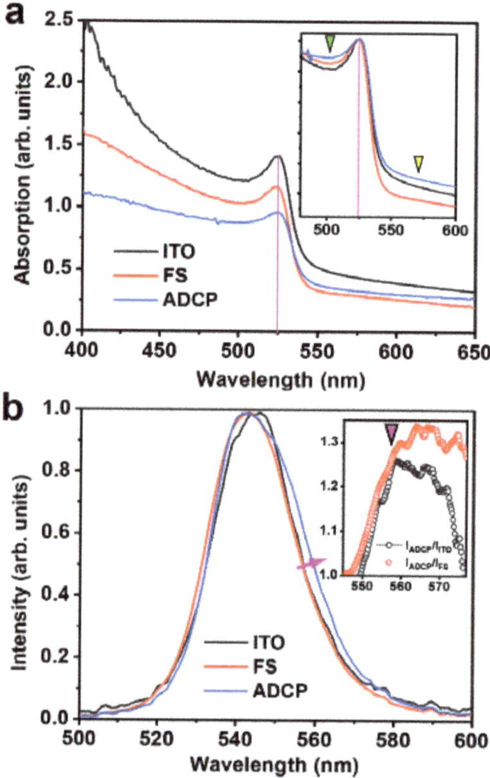

Figure 2. (**a**) Absorption spectra of $MAPbBr_3$ coated on substrates of ADCP, ITO, and FS. Inset: normalized comparison in a selective spectral range. (**b**) PL spectra of $MAPbBr_3$ on different substrates. Inset: calculated spectra by $I_{ADCP}(\lambda)/I_{ITO}(\lambda)$ and $I_{ADCP}(\lambda)/I_{FS}(\lambda)$.

4. Random Lasing Properties on Different Substrates

In the investigation on the random lasers, we used 150-fs laser pulses at 400 nm as the pump; these are produced by frequency-doubling of 800-nm pulses from a Ti:sapphire amplifier and have a repetition rate of 1000 Hz. An optical attenuator was placed between the pump laser and the samples of $MAPbBr_3$-coated substrates, so that the pump fluence could be adjusted continuously. A fiber spectrometer was mounted in front of the front surface of the sample so that the detection avoids the reflection of the pump laser beam. Random lasing spectra are measured at different pump fluences, as shown in Figure 3.

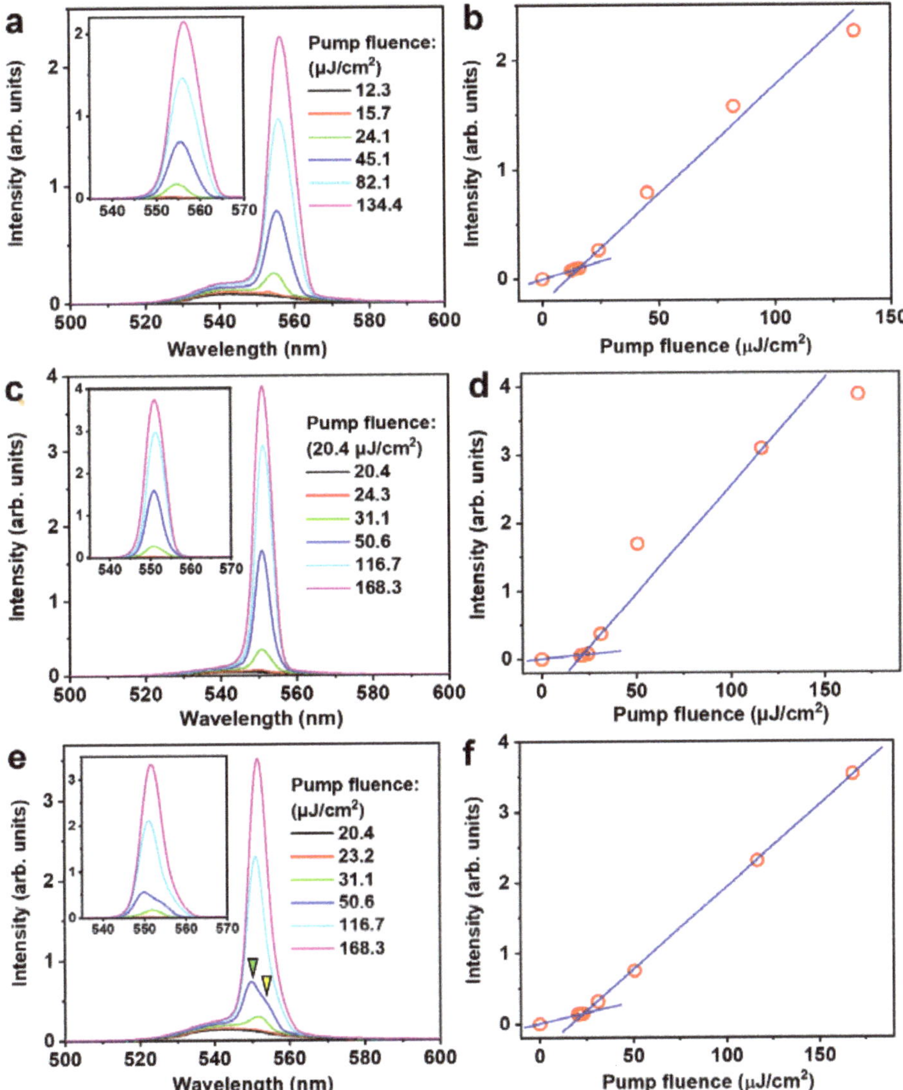

Figure 3. Random lasing properties measured on $MAPbBr_3$ coated on substrates of (**a**,**b**) ADCP, (**c**,**d**) ITO, and (**e**,**f**) FS. (**a**,**c**,**e**) Directly measured random lasing spectra at different pump fluences. (**b**,**d**,**f**) Pump threshold properties corresponding to substrates ADCP, ITO, and FS, respectively. Insets: Pure random lasing spectra calculated by subtracting the background PL spectrum from the directly measured radiation spectra.

Figure 3a shows the random lasing spectra measured on the sample with an ADCP substrate as the pump fluence was increased from 12.3 to 134.4 μJ/cm^2. Due to the large variety of the random structures, relatively broader lasing spectra can be observed, where a bandwidth of 7.7 nm at FWHM was measured for a pump fluence of 134.4 μJ/cm^2, as shown in the Figure 3a inset. The lasing spectra in the Figure 3a inset have been processed by subtracting the background PL spectra before the random lasing action; therefore, the data in the inset are in fact pure random lasing spectra, where the increase of the pump fluence starts at 15.7 μJ/cm^2. Figure 3b plots the random lasing intensity as a function of pump fluence, where two distinct stages can be observed, justifying a pump threshold between 12.3 and 15.7 μJ/cm^2.

In Figure 3c,d, we present the experiment results of the samples with an ITO substrate, where we show the pump fluence dependence of the random lasing spectrum and the peak intensity, respectively. From the measurement results, we may measure a peak wavelength of 551.1 nm and a narrower bandwidth of 5.5 nm at FWHM, as shown in the inset in Figure 3c, with respect to the case for an ADCP substrate. However, the pump fluence was found to have increased to between 20.4 and 24.3 μJ/cm^2. We may thus understand that large crystal particles in the ADCP-substrate structures led to broadened lasing spectra but lowered pump threshold, where the larger range of the crystal-size distribution facilitates stronger optical scattering in a broader spectrum.

Similar pump threshold performance can be observed with random lasing performance for an FS substrate, which may be estimated between 20.4 and 23.2 μJ/cm^2, according to Figure 3e,f. Meanwhile, as shown in the Figure 3e inset, we measured a spectral bandwidth of about 5.97 nm at FWHM for a pump fluence of 168.3 μJ/cm^2. However, for the FS substrate we may observe competition between two lasing modes, which can be clearly identified when the pump fluence is lower than 50.6 μJ/cm^2. These two lasing modes are located at 551.7 and 554 nm, as highlighted by the green and yellow triangles in Figure 3e, respectively. This feature can be more clearly seen in Figure S3, where we fitted the lasing spectrum at a pump fluence of 50.6 μJ/cm^2 using Gaussian functions, so that we were able to decompose the spectral structures. We can clearly see the locations of the PL spectrum (green) centered at 545.2 nm and the two lasing bands peaking at 551.7 (red) and 554 nm (magenta). Although the crystal particles are even smaller on the FS substrate than on the ITO, the wrinkling modulation becomes much larger on the FS substrate, as highlighted by the dashed circles in Figure 1(c_2,c_3). The occurrence of a second lasing mode may thus be explained by the much-enhanced surface morphology modulation on the FS-substrate, which can be identified by the larger protruding wrinkles than on the other two substrates. The strong scattering by larger centers corresponds to the enhanced gain at longer wavelengths during the interaction between the emissions and excited molecules.

To demonstrate a better comparison between the lasing performance of the samples on three different substrates, we replotted the normalized random lasing spectra at varied pump fluences for the three samples in Figure S4, where we fixed a spectral range of 18 nm for these three groups of data, so that both the tuning range and the bandwidth of the lasing spectrum were comparable for the three samples. We can clearly see that the largest redshift of the lasing spectrum when increasing the pump fluence is produced for the ADCP-substrate sample, as indicated by the red arrow toward the right in Figure S4a. Although a relatively small redshift is observed with the ITO-substrate, there is a definite monotonic redshift (the smaller red arrow toward the right in Figure S3b) of the lasing spectrum when increasing the pump fluence. However, no monotonic shift of the lasing spectrum is observed for the FS substrate when increasing the pump fluence, as highlighted by a double arrow in Figure S4c. This is because there are two lasing modes interacting with each other for the FS substrate. Furthermore, the broadest lasing spectrum is observed for the ADCP substrate, which agrees very well with the microscopic properties of the $MAPbBr_3$ film on ADCP. Large crystal particles, as well as high-density pin-hole and crack defects, have broadened the excitation bands and, consequently, also broadened the lasing spectrum. These defects may induce subbands below the excitation bandedge of the molecules, resulting in new emission features with a red-shifted spectrum. Meanwhile, the consequently stronger optical scattering mechanisms led to a much-reduced lasing threshold for the ADCP substrate, as compared with the other two.

In Figure 4, we make a more detailed comparison between the lasing spectra measured on $MAPbBr_3$ on ADCP, ITO, and FS substrates for pump fluences of 134.4, 168.3, and 168.3 μJ/cm^2, respectively. The corresponding peak wavelengths are located at 551.1, 551.7, and 556.5 nm, respectively. The ADCP substrate supports an almost completely separate lasing spectrum from the ITO and FS substrates, which is also the broadest among the three. Although the samples based on the ITO and FS substrates exhibit a quite similar surface morphology, the FS substrate supports a slightly red-shifted and broadened lasing spectrum. More importantly, for the FS substrate, we can still identify an additional feature on the falling edge of the lasing spectrum, as highlighted by a dashed circle, which agrees well with our discussions above regarding the mode competition effect. Therefore, the surface morphological performance of the spin-coated $MAPbBr_3$ film, as well as its spectroscopic response, is very sensitive to the material properties of the substrate.

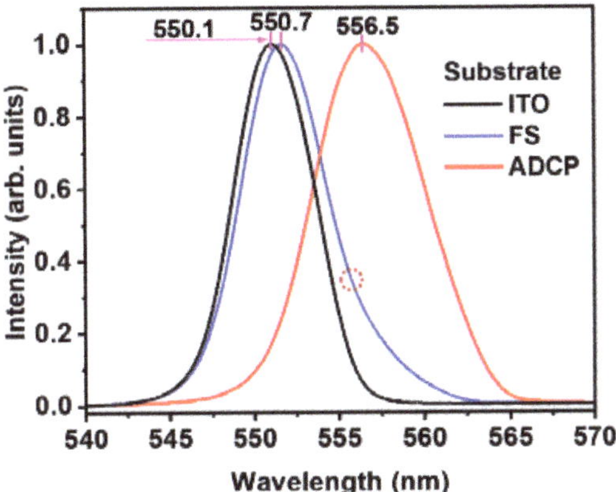

Figure 4. Comparison between lasing spectra from $MAPbBr_3$ coated on the ITO (black), FS (blue), and ADCP (red) substrates.

Furthermore, we may consider the stronger defect sites produced in the ADCP- and FS-substrate samples than those in the ITO-substrate, as shown in Figure 1. The defect states with lowered energy levels in the excitation band are responsible for the new emission features with a red-shifted spectrum. This reasonably explains the red-shifted and broadened random lasing spectrum for the ADCP-substrate sample, as well as a second spectral feature at a longer wavelength for the FS-substrate sample, as has been clearly depicted in Figure S3.

5. Materials and Methods

5.1. Chemicals

Methylamine Bromide (MABr) and Lead (II) Bromide (PbBr2) were purchased from Xi'an Polymer Light Technology Corp., N,N-dimethylformamide (DMF, 99.8%) from Adamas, Tricyclo[5.2.1.$0^{2,6}$] decanedimethanol diacrylate (A-DCP) from SHIN-NAKAMURA CHEMICAL CO, LTD. (Wakayama, Japan), dipentaerythritol penta-/hexa-acrylate from Shanghai Aladdin Biochemical Technology Co., Ltd. (Shanghai, China), and 2,2–Dimethoxy-2-phenylacetophenone (DMPA) from Shanghai Macklin Biochemical Co., Ltd. (Shanghai, China). All regents were used directly without further purification.

5.2. Preparation of the ADCP Substrate

Wash a FS substrate with a size of 15 $\times$ 15mm^2 and a thickness of 1 mm by TFD 7, deionized water, acetone, and absolute ethanol sequentially for 30 min. Next, blow it with clean nitrogen to dry it. Put the A-DCP, acrylic, and DMPA in a 3 mL bottle to mix them and stir them for 2 h under red light to obtain the ADCP precursor solution. Following this, drop the ADCP solution onto the cleaned FS substrate with a volume of about 10 μL and cover it with another cleaned FS substrate, and clamp together tightly. Finally, place the sandwich structure under the UV lamp for 1 min to complete curing of the ADCP. Remove the FS substrate on the top and finish the preparation of the ADCP substrate on the bottom part.

5.3. Preparation of the ITO and FS Substrates

Wash the ITO or FS substrate by TFD 7, deionized water, acetone, and absolute ethanol sequentially for 30 min. Both types of substrates have a dimension of 15 mm × 15 mm × 1 mm. Blow the cleaned glass substrate with clean nitrogen, until it is dried.

5.4. Synthesis of MAPbBr3

Dissolve equal molar mass of methylamine bromide (MABr) and lead bromide (PbBr2) in 1 mL N,N-dimethylformamide (DMF), heat the mixture at 60 °C and stir it overnight to ensure they are fully dissolved and mixed. The MAPbBr3 solution is thus obtained with a concentration of 0.56 g/mL.

5.5. Preparation of the MAPbBr3 Film on Different Substrates

Spin-coat the solution of MAPbBr3 in DMF with a concentration of 0.56 g/mL onto the ITO, FS, and ADCP substrates in a glove box. The spin-coating speed is 3000 rpm and the duration 30 s, which is the same for all of the samples. Then, place the spin-coated samples on a hotplate and heat them at 80 °C for about 1 min to evaporate the remaining solvent.

5.6. Contact Angle Measurements

Put the cleaned ITO, FS, and ADCP substrates on the sample table of the contact angle measurement instrument from Shanghai FangRui Instrument Co., Ltd. (Shanghai, China), respectively. Drop 2 μL MAPbBr3 solution with a concentration of 0.56 g/mL on each substrate using a microinjector and measure the contact angle using the instrument-equipped software. Capture the displayed picture with the measurement result of the contact angles.

5.7. Measurements on Random Laser Performance

The pump laser source is supplied by a frequency-doubled output from a Ti:sapphire amplifier, which is centered at 400 nm, having a pulse length of about 150 fs and a repetition rate of 1 KHz. The emission spectrum is measured using the Maya2000 Pro fiber spectrometer from Ocean Optics (Dunedin, US), which has a spectral resolution of 0.1 nm.

6. Conclusions

We investigated the dependence of the surface morphology, absorption and emission spectra, and the random lasing properties of the spin-coated organic–inorganic hybrid perovskite $MAPbBr_3$ on the substrate materials. Making using of the efficient photoluminescence properties of $MAPbBr_3$, we were able to achieve random lasing emissions easily in various devices fabricated by direct spin-coating. The bandwidth and location of the random lasing spectrum are found to be tuned by the size of the crystallization particles and by the surface morphologies of the multicrystalline film, which is determined sensitively by the wetting properties of the solution of $MAPbBr_3$ in DMF on the substrates during the spin-coating process. The revealed mechanisms are important for the design and application of solution-processed thin-film laser devices.

Supplementary Materials: The following supporting information can be downloaded at: https://www.mdpi.com/article/10.3390/cryst12030334/s1, Figure S1: Optical microscopic image of an ADCP grating with a period of 4 μm before (a) and after (b) being immersed in DMF; Figure S2: PL spectra of the pure FS, ITO, and ADCP substrates under the excitation of 150 fs pulses at 400 nm. Inset: enlarged view in the studied spectral range from 540 to 560 nm. Only noise is observed in the PL spectra without any typical spectral features; Figure S3: The measured emission spectrum of the ADCP-substrate $MAPbBr_3$ at a pump fluence of 50.6 μJ/cm^2 of 400-nm excitation pulses and the multipeak fittings to resolve two distinct random lasing bands peaked at 547.1 and 554 nm by the red and magenta curves, respectively; Figure S4: Pump fluence dependence of the normalized random lasing spectrum for the (a) ADCP, (b) ITO, and (c) FS substrates.

Author Contributions: Author Contribution: Conceptualization, X.Z.; Formal analysis, X.Z.; Funding acquisition, X.Z.; Investigation, J.H.; Methodology, X.Z.; Project administration, X.Z.; Resources, X.Z.; Supervision, X.Z.; Validation, X.Z.; Visualization, X.Z.; Writing—original draft, X.Z.; Writing—review & editing, X.Z. All authors have read and agreed to the published version of the manuscript.

Funding: This research received no external funding.

Acknowledgments: The authors acknowledge the National Natural Science Foundation of China (61735002, 12074020) and Beijing Municipal Education Commission (KZ202010005002) for the support.

Conflicts of Interest: The authors declare no conflict of interest.

References

1. Chen, S.S.; Xiao, X.; Chen, B.; Kelly, L.L.; Zhao, J.J.; Lin, Y.Z.; Toney, M.F.; Huang, J.S. Crystallization in One-Step Solution Deposition of Perovskite Films: Upward or Downward? *Sci. Adv.* **2021**, *7*, 2412. [CrossRef]
2. Kim, Y.H.; Kim, S.; Kakekhani, A.; Park, J.; Lee, Y.H.; Xu, H.X.; Nagane, S.; Wexler, R.B.; Kim, D.H.; Jo, S.H.; et al. Comprehensive Defect Suppression in Perovskite Nanocrystals for High-efficiency Light-Emitting Diodes. *Nat. Photonics* **2021**, *15*, 148. [CrossRef]
3. Zhang, Q.; Shang, Q.Y.; Su, R.; Do, T.T.H.; Xiong, Q.H. Halide Perovskite Semiconductor Lasers: Materials, Cavity Design and Low Threshold. *Nano Lett.* **2021**, *21*, 1903. [CrossRef] [PubMed]
4. Liu, Z.Z.; Huang, S.H.; Du, J.; Wang, C.W.; Leng, Y.X. Advances in Inorganic and Hybrid Perovskites for Miniaturized Lasers. *Nanophotonics* **2020**, *9*, 2251. [CrossRef]
5. Hu, J.Y.; Xue, H.B.; Zhang, X.P. Two-Dimensional Crystalline Gridding Networks of Hybrid Halide Perovskite for Random Lasing. *Crystals* **2021**, *11*, 1114. [CrossRef]
6. Hassan, A.; Azam, M.; Ahn, Y.H.; Zubair, M.; Cao, Y.; Khan, A.A. Low Dark Current and Performance Enhanced Perovskite Photodetector by Graphene Oxide as an Interfacial Layer. *Nanomaterials* **2022**, *12*, 190. [CrossRef]
7. Kao, T.S.; Hong, Y.H.; Hong, K.B.; Lu, T.C. Perovskite Random Lasers: A Tunable Coherent Light Source for Emerging Applications. *Nanotechnology* **2021**, *32*, 282001. [CrossRef]
8. Hu, J.Y.; Wang, M.; Tang, F.W.; Liu, M.; Mu, Y.Y.; Fu, Y.L.; Guo, J.X.; Song, X.Y.; Zhang, X.P. Threshold Size Effects in the Patterned Crystallization of Hybrid Halide Perovskites for Random Lasing. *Adv. Photonics Res.* **2021**, *2*, 2000097. [CrossRef]
9. Liu, Y.L.; Yang, W.H.; Xiao, S.M.; Zhang, N.; Fan, Y.B.; Qu, G.Y.; Song, H.Q. Surface-Emitting Perovskite Random Lasers for Speckle-Free Imaging. *ACS Nano* **2019**, *13*, 10653. [CrossRef] [PubMed]
10. Hong, Y.H.; Kao, T.S. Room-Temperature Random Lasing of Metal-Halide Perovskites via Morphology-Controlled Synthesis. *Nanoscale Adv.* **2020**, *2*, 5833. [CrossRef]
11. Wang, C.C.; Kataria, M.; Lin, H.; Nain, A.; Lin, H.Y.; Inbaraj, C.R.; Liao, Y.M.; Thakran, A.; Chang, H.T.; Tseng, F.G.; et al. Generation of Silver Metal Nanocluster Random Lasing. *ACS Photonics* **2021**, *8*, 3051. [CrossRef]
12. Sato, R.; Henzie, J.; Zhang, B.Y.; Ishii, S.; Murai, S.; Takazawa, K.; Takeda, Y. Random Lasing via Plasmon-Induced Cavitation of Microbubbles. *Nano Lett.* **2021**, *21*, 6064. [CrossRef] [PubMed]
13. Zhu, H.Y.; Zhang, W.L.; Zhang, J.C.; Ma, R.; Wang, Z.; Rao, Y.J.; Li, X.F. Single-Shot Interaction and Synchronization of Random Microcavity Lasers. *Adv. Mater. Technol.* **2021**, *6*, 2100562. [CrossRef]
14. Ejbarah, R.A.; Jassim, J.M.; Haddawi, S.F.; Hamidi, S.M. Transition from Incoherent to Coherent Random Lasing by Adjusting Silver Nanowires. *Appl. Phys. A* **2021**, *127*, 476. [CrossRef]
15. Gayathri, R.; Monika, K.; Murukeshan, V.M.; Vijayan, C. Low Threshold Incoherent Random Lasing with Spectral Overlap Optimization of Size-Tuned Plasmonic Nanorods. *Opt. Laser Technol.* **2021**, *139*, 106959. [CrossRef]
16. Okamoto, T.; Mori, M. Random Laser Action in Dye-Doped Polymer Media with Inhomogeneously Distributed Particles and Gain. *Appl. Sci.* **2019**, *9*, 3499. [CrossRef]
17. Yang, T.H.; Chen, C.W.; Jau, H.C.; Feng, T.M.; Wu, C.W.; Wang, C.T.; Lin, T.H. Liquid-crystal random fiber laser for speckle-free imaging. *Appl. Phys. Lett.* **2019**, *114*, 191105. [CrossRef]
18. Giorgi, M.L.; Lippolis, T.; Jamaludin, N.F.; Soci, C.; Bruno, A.; Anni, M. Origin of Amplified Spontaneous Emission Degradation in $MAPbBr_3$ Thin Films under Nanosecond-UV Laser Irradiation. *J. Phys. Chem. C* **2020**, *124*, 10696. [CrossRef]
19. Murzin, A.O.; Stroganov, B.V.; Günnemann, C.; Hammouda, S.B.; Shurukhina, A.V.; Lozhkin, M.S.; Emeline, A.V.; Kapitonov, Y.V. Amplified Spontaneous Emission and Random Lasing in $MAPbBr_3$ Halide Perovskite Single Crystals. *Adv. Optical Mater.* **2020**, *8*, 2000690. [CrossRef]
20. Pourdavoud, N.; Mayer, A.; Buchmüller, M.; Brinkmann, K.; Häger, T.; Hu, T.; Heiderhoff, R.; Shutsko, I.; Görrn, P.; Chen, Y.W.; et al. Distributed Feedback Lasers Based on $MAPbBr_3$. *Adv. Mater. Technol.* **2018**, *3*, 1700253. [CrossRef]
21. Chen, R.X.; Su, X.Q.; Wang, J.; Gao, D.W.; Pan, Y.; Wang, Y.M.; Wang, L. The Roles of Surface Defects in $MAPbBr_3$ and Multi-Structures in $MAPbI_3$. *Opt. Mater.* **2021**, *122*, 111600. [CrossRef]
22. Khan, A.A.; Yu, Z.N.; Khan, U.; Dong, L. Solution Processed Trilayer Structure for High-Performance Perovskite Photodetector. *Nanoscale Res. Lett.* **2018**, *13*, 399. [CrossRef] [PubMed]

23. Ahmad, R.; Surendran, A.; Harikesh, P.C.; Haselsberger, R.; Jamaludin, N.F.; John, R.A.; Koh, T.M.; Bruno, A.; Leong, W.L.; Mathews, N.; et al. Perturbation-Induced Seeding and Crystallization of Hybrid Perovskites over Surface-Modified Substrates for Optoelectronic Devices. *ACS Appl. Mater. Interfaces* **2019**, *11*, 27727. [CrossRef] [PubMed]
24. Gu, Z.K.; Zhou, Z.H.; Huang, Z.D.; Wang, K.; Cai, Z.R.; Hu, X.T.; Li, L.H.; Li, M.Z.; Zhao, Y.S.; Song, Y.L. Controllable Growth of High-Quality Inorganic Perovskite Microplate Arrays for Functional Optoelectronics. *Adv. Mater.* **2020**, *3*, 1908006. [CrossRef]
25. Makarov, S.; Furasova, A.; Tiguntseva, E.; Hemmetter, A.; Berestennikov, A.; Pushkarev, A.; Zakhidov, A.; Kivshar, Y. Halide-Perovskite Resonant Nanophotonics. *Adv. Opt. Mater.* **2019**, *7*, 1800784. [CrossRef]
26. Berestennikov, A.S.; Voroshilov, P.M.; Makarov, S.V.; Kivshar, Y.S. Active Meta-Optics and Nanophotonics with Halide Perovskites. *Appl. Phys. Rev.* **2019**, *6*, 031307. [CrossRef]

MDPI

Article

Narrowband Near-Infrared Perovskite/Organic Photodetector: TCAD Numerical Simulation

Marwa S. Salem [1,2], Ahmed Shaker [3,*], Amal H. Al-Bagawia [4], Ghada Mohamed Aleid [5], Mohamed S. Othman [5], Mohammad T. Alshammari [6] and Mostafa Fedawy [7,8]

1 Department of Computer Engineering, College of Computer Science and Engineering, University of Hail, Ha'il 55211, Saudi Arabia; marwa_asu@yahoo.com
2 Department of Electrical Communication and Electronics Systems Engineering, Faculty of Engineering, Modern Science and Arts University (MSA), Cairo 12556, Egypt
3 Engineering Physics and Mathematics Department, Faculty of Engineering, Ain Shams University, Cairo 11566, Egypt
4 Chemistry Department, Faculty of Science, University of Ha'il, Hail 55211, Saudi Arabia; a.albagawi@uoh.edu.sa
5 B.Sc. Department, Preparatory Year College, University of Ha'il, Hail 55211, Saudi Arabia; g.aleid@uoh.edu.sa (G.M.A.); mo.abdelkarim@uoh.edu.sa (M.S.O.)
6 Department of Computer Science and Information, Computer Science and Engineering College, University of Ha'il, Hail 55211, Saudi Arabia; m.alsagri@uoh.edu.sa
7 Electronics and Communications Department, Faculty of Engineering, Arab Academy for Science and Technology and Maritime Transport, Cairo 11736, Egypt; m.fedawy@aast.edu
8 Center of Excellence in Nanotechnology, Arab Academy for Science and Technology and Maritime Transport, Cairo 11736, Egypt
* Correspondence: ahmed.shaker@eng.asu.edu.eg

Citation: Salem, M.S.; Shaker, A.; Al-Bagawia, A.H.; Aleid, G.M.; Othman, M.S.; Alshammari, M.T.; Fedawy, M. Narrowband Near-Infrared Perovskite/Organic Photodetector: TCAD Numerical Simulation. *Crystals* **2022**, *12*, 1033. https://doi.org/10.3390/cryst12081033

Academic Editors: Xinping Zhang, Baoquan Sun and Fujun Zhang

Received: 6 July 2022
Accepted: 21 July 2022
Published: 25 July 2022

Publisher's Note: MDPI stays neutral with regard to jurisdictional claims in published maps and institutional affiliations.

Abstract: Narrowband photodetectors (PD) established in the near-infrared (NIR) wavelength range are highly required in a variety of applications including high-quality bioimaging. In this simulation study, we propose a filter-less narrowband PD based on the architecture of perovskite/organic heterojunction. The most decisive part of the photodetector is the hierarchical configuration of a larger bandgap perovskite material with a thicker film followed by a lower bandgap organic material with a narrower layer. The design of the structure is carried out by TCAD numerical simulations. Our structure is based on an experimentally validated wideband organic PD, which is modified by invoking an additional perovskite layer having a tunable bandgap. The main detector device comprises of ITO/perovskite ($Cs_yFA_{1-y}Pb(I_xBr_{1-x})_3$)/organic blend (PBDTTT-c:C60-PCBM)/PEDOT:PSS/Al. The simulation results show that the proposed heterojunction PD achieves satisfactory performance when the thickness of perovskite and organic layers are 2.5 μm and 500 nm, respectively. The designed photodetector achieves a narrow spectral response at 730 nm with a full width at half-maximum (FWHM) of 33 nm in the detector, while having a responsivity of about 0.12 A/W at zero bias. The presented heterojunction perovskite/organic PD can efficiently detect light in the wavelength range of 700 to 900 nm. These simulation results can be employed to drive the development of filter-less narrowband NIR heterojunction PD.

Keywords: narrowband; near-infrared; perovskite; organic; TCAD; responsivity; FWHM

1. Introduction

Narrowband photodetectors (PDs) have been broadly utilized in various application fields like biomedical imaging, virtual reality, navigation aid, full-weather robots, and many others [1–5]. In these PDs, light can be detected inside a specific wavelength range, and there is no light response at other wavelengths. The existing commercial market of PDs is dominated by expensive crystalline inorganic semiconductor structures, which are usually integrated with optical filters [6]. Meanwhile, organic semiconductors can compete the existing technologies because they are lightweight and flexible besides their low-cost and semi-transparent nature. In addition, organic materials have high absorption coefficients and can effectively be designed as absorber materials at thickness lower than only 1 μm. Additionally, the absorption spectrum of organic materials can be modulated by modifying their molecular configurations. These promising features make organic PDs prospective contenders for the expanding need for smarter and safer detectors [7–9].

Most organic PDs are composed of a blend of a polymer donor and fullerene acceptor molecules as the photoactive thin film [10]. Although broadband organic PDs can be commonly attained [11–13], it is not that easy to achieve narrowband organic PDs due to the relatively wide photon harvesting range of organic semiconductors [14,15]. Moreover, organic–inorganic hybrid perovskite materials are developing candidates that have been extensively used in PDs [16]. Some of the advantages of perovskites are their pronouncing optical and electrical properties, involving a direct bandgap, large extinction coefficient, and high carrier mobility [17,18]. It has been shown that by combining organic–inorganic hybrid perovskites and an organic heterojunction comprising of donor–acceptor materials can push this type of PDs to cover the near-infrared range resulting in broadband PDs [19,20]. A hybrid PD with perovskite/polymer, which is based on $CH_3NH_3PbI_{3-x}Br_x$/PTB7 with tunable spectral response in the range 680–710 nm, has been reported [21]. Furthermore, it has been demonstrated that a hybrid PD with perovskite/polymer heterojunction based on $CH_3NH_3PbI_3$/PCPDTBT:PC_{71}BM can achieve a visible-blind narrowband NIR detection [22].

Notably, the biological tissues are primarily comprised of hemoglobin and water that absorb wavelengths below 650 nm and above 900 nm [23]. Consequently, in order to prevent the light absorption of biological tissue, a PD operating in the range 650 to 900 nm is required [24]. Therefore, our aim in this work is to design a narrowband PD that is based on perovskite/organic heterojunction to provide an optimal detection region. The design idea is based on selecting a relatively wide band gap perovskite that can fully absorb the incident photons whose wavelengths are smaller than the perovskite cutoff wavelength [25]. This can be accomplished if the perovskite layer has a wider thickness and higher defect density than the organic layer [26]. When the light passes through the perovskite, the visible light is absorbed while the NIR light will be transferred to the organic layer due to the transparency of the perovskite layer in the NIR region. So, the organic material will respond to the light spectrum between the cutoff wavelengths corresponding to the organic and perovskite materials. Next, the photoexcited charge carriers in the organic layer will be separated by the produced built-in electric field, which is a process that is nearly not impacted by the defects in the organic film. This perovskite/organic PD design does not require a complex filtering system and may operate under zero bias due to the presence of a built-in electric field at the heterojunction. So, it can be used as a filter-free and self-powered device.

In order to design and reveal the internal physics of an optoelectronic device, a TCAD simulation meticulously addressing the basic optical, electrical, and semiconductor characteristics is enormously valuable. Such an advanced simulation can accomplish an accurate performance evaluation of various PDs and provide exact physical pictures that reflect the detailed device operation. Compared to the complex and time-consuming experiments, a highly effective and precise simulation can be extremely beneficial for predicting and optimizing the device performance in a convenient way. In this work, numerical simulations were carried out using Silvaco TCAD device simulator package [27]. We performed finite element simulations enabling electro-optical modeling of our proposed PD. Currently, such a simulation of narrowband perovskite/organic PD has never been reported.

This work thus focuses on the design and numerical analysis of a narrowband perovskite/organic PD. The selected perovskite material is $Cs_yFA_{1-y}Pb(I_xBr_{1-x})_3$ whose bandgap can be tuned in the range 1.5 to 1.8 eV [28]. Meanwhile, the organic material is a blend between the donor polymer PBDTTT-c and the acceptor C60-PCBM. This organic material has a band gap of about 1.45 eV, and PDs based on this organic semiconductor have been experimentally validated as wideband PDs [29]. Thus, a calibration step will be presented to confirm the physical and geometrical parameters used in simulation by comparing the TCAD simulation results versus the experimental data [29]. Then, we will study the impact of the energy gap of the perovskite material to get the optimum choice based on responsivity (*R*) and full width at half-maximum (FHWM) parameters. Moreover, the impact of reverse bias and organic thickness and trap density is investigated.

2. Simulation Methodology and Device Model

To build our structure model and simulate the electrical and optical characteristics of the proposed PD, the Silvaco TCAD device simulator module is utilized. The involved numerical simulation is based on the discretized solution of the basic equations of charge transport in semiconductors like Poisson's equation, continuity, and transport equations. Firstly, given an illuminated input source, the optical intensity profiles within the PD device are calculated. Then, the intensity profiles are transformed into photogeneration rates, which are integrated into the generation terms in the continuity equations. In optoelectronic device simulation, two separate models are computed concurrently at each bias. These models are the optical ray tracing and the photogeneration model. In the first model, the real component of the refractive index is adopted to evaluate the optical intensity, while, in the second model, the extinction coefficient is employed in the calculation of a new carrier concentration. An electrical simulation is then accomplished to acquire the required terminal currents [30], where the drift diffusion model is employed to simulate the transport properties. The physical models used in simulation are as follows. The Shockley–Read–Hall (SRH) recombination model and Poole–Frenkel mobility model were enabled. Notably, the SRH recombination mechanism arises from the recombination of electron–hole pairs through defect levels within the energy bandgap of the simulated material [31,32].

2.1. Basic Photodetector Structure

The schematic diagram of the proposed detector is displayed in Figure 1a. In the simulation process, the perovskite/organic heterojunction PD structure was generated by the device simulator on a 2D grid as displayed in Figure 2b. The corresponding energy levels of the distinct layers are displayed in Figure 1c, while the energy band diagram at dark condition is plotted in Figure 1d. As depicted in the schematic figure, the basic structure of the PD is ITO/PEIE followed by $Cs_yFA_{1-y}Pb(I_xBr_{1-x})_3$. This perovskite material was chosen because it is more stable than MA-based compounds [33] and due to its bandgap tunability [28]. Thus, $Cs_yFA_{1-y}PbI_xBr_{(1-x)3}$ can serve as a suitable partner in the heterojunction detector combining with a proper organic material. Light is incident from the ITO side and passes through the perovskite film where photons of wavelengths smaller than a certain designed wavelength are suppressed (see Figure 1c). The active organic film is PBDTTT-c:C60-PCBM (donor/acceptor). The thickness of the perovskite and organic layers is set as 2.5 μm and 500 nm, respectively. The hole transport layer (HTL) is PEDOT:PSS with a p-doping of about 1×10^{18} cm^{-3}.

A summary of the technological and physical parameters of the PD layers is presented in Table 1 [28,29,34–39]. Further, Table 2 lists the main defect parameters inside the perovskite and organic blend layers [24,29]. The details and criteria of choosing these parameters are explained hereafter. Moreover, the refractive indices and distinction coefficients, which are extracted from experimental reports, are displayed in the Supplementary Materials [28,29,37]. It should be pointed out here that our proposed PD structure is based on a fabricated wideband organic PD. The modification made here is to add a thick perovskite layer in order to engineer the band of detection. Before proceeding to present our PD results, a validation of the parameters of the materials used in our PD design is conducted. This is done by comparing TCAD simulation results with those obtained from measurements of the fabricated organic PD [29]. The Silvaco script of this part is listed in the Supplementary Materials.

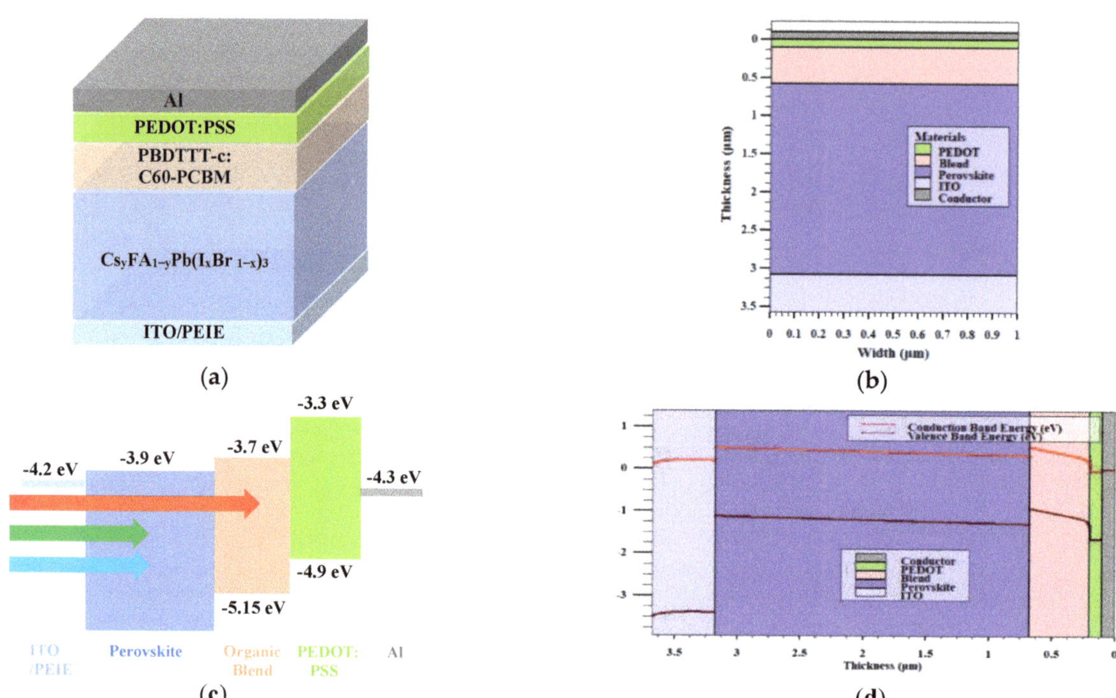

Figure 1. Basic structure and energy levels of the proposed narrowband near-infrared perovskite/organic photodetector. (**a**) Schematic illustration of the structure of PDs, (**b**) generated structure from the device simulator, (**c**) energy level diagram of the different layers, and (**d**) energy band diagram at dark condition showing conduction and valence edges.

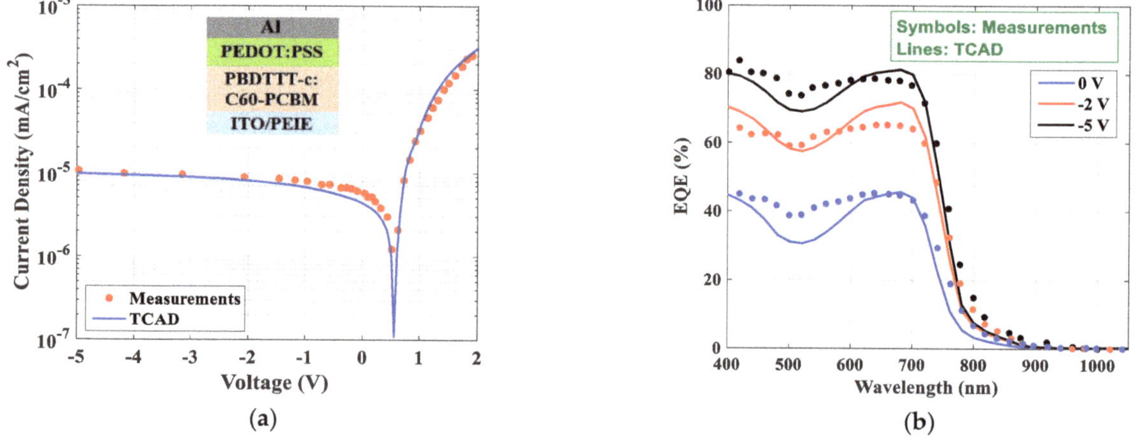

Figure 2. Calibration of experimental wideband organic PD (**a**) illuminated *J*-*V* characteristics (530 nm and 0.3 W/m^2). The main structure of the PD is shown in the inset and (**b**) EQE characteristics.

Table 1. Basic parameters of the narrowband PD layers [28,29,34–39].

Parameters	ITO/PEIE	$Cs_yFA_{1-y}Pb(I_xBr_{1-x})_3$	PBDTTT-c: C60-PCBM	PEDOT:PSS
Thickness (nm)	110	2500	500	100
Energy gap (eV)	3.60	1.62–1.80	1.45	1.60
Electron affinity (eV)	4.20	3.90	3.70	3.30
Relative permittivity	9.0	7.0	3.6	3.0
Electron mobility ($cm^2/V.s$)	100	10	6×10^{-4}	5×10^{-4}
Hole mobility ($cm^2/V.s$)	25	10	1×10^{-3}	5×10^{-4}
CB effective density of states (cm^{-3})	2.2×10^{18}	2.75×10^{18}	1.0×10^{20}	2.2×10^{18}
VB effective density of states (cm^{-3})	1.8×10^{19}	3.90×10^{18}	1.0×10^{20}	1.8×10^{19}

Table 2. Bulk defects parameters in the perovskite and organic layers [24,29].

	$Cs_yFA_{1-y}Pb(I_xBr_{1-x})_3$	PBDTTT-c: C60-PCBM	PBDTTT-c:C60-PCBM
Defect type	Donor	Donor	Neutral
Electron and hole capture cross section	1×10^{-13} cm^2	5×10^{-16} cm^2	1×10^{-15} cm^2
Trap energy position	Mid-gap	290 meV (Above blend HOMO)	Mid-gap
Total density (N_t)	1×10^{15} cm^{-3}	2.7×10^{16} cm^{-3}	2×10^{14} cm^{-3}

2.2. Calibration of the Fabricated Wideband Organic PD

The experimental device, used in calibration, consists of multiple layers stacked as shown in the inset of Figure 2a. The transparent conductive oxide is ITO/Polyethylenimide (PEIE), which is utilized to reduce the overall work function. The thickness of ITO/PEIE is 110 nm and is followed by the active layer. The active layer is PBDTTT-c (as donor material) and C60-PCBM (as acceptor material) with a 1:1.5 ratio in weight, and its thickness is 500 $\pm$ 20 nm. PEDOT:PSS, as HTL, is deposited on top of the active layer. The last deposited layer is aluminum, whose thickness is 100 nm. The work functions of ITO/PEIE and PEDOT:PSS are 4.2 and 4.9 eV, respectively, as determined by Kelvin probe [29]. The optical parameters of the distinct layers, namely, the refractive index and extinction coefficient, are taken from [39]. Regarding the active material physical parameters, they are extracted from the literature [29,38]. The LUMO and HOMO levels are 3.7 and 5.15 eV, respectively, giving a blend bandgap of 1.45 eV. Further, the electron and hole mobility is taken as 6×10^{-4} and 1×10^{-3} cm^2/V.s, respectively [38]. Regarding the bulk trap defects, a donor trap density is estimated to be 2.7×10^{16} cm^{-3} at a position of 290 meV above the blend HOMO, while a trap density of 2×10^{14} cm^{-3} of a mid-gap trap state is fitted [39].

Figure 2a shows the current density vs. voltage of this wideband detector that was measured under 530 nm LED light having an intensity of 0.3 W/m^2. The figure also shows the TCAD simulation results, which are fitted by adjusting the conduction and valence density of states and trap densities. More details about the calibration are found in [29,39]. Furthermore, Figure 2b exhibits the EQE characteristics under different bias conditions for the experimental and simulation results. The good agreement between the simulation results and those from measurements indicates a satisfactory validation of the material parameters and physical models applied in the simulator.

3. Results and Discussions

In this section, the TCAD simulation results of our proposed perovskite/organic material system PD are presented. The influence of trap density and thickness of the perovskite film is studied to get a design guideline about the main parameters of the perovskite layer that assure a narrowband detector operation. Next, the effect of the variation of the perovskite energy gap is explored. Finally, the impact of reverse bias on the PD main parameters is investigated. The key factors of the PD under study are extracted from the simulator once the simulation process is done. Besides the dark and illuminated output current, other parameters like responsivity (R) and full width at half maximum (FWHM) are considered significant metrics that can measure the effectiveness of the detector and also can be used to differentiate between different types of PDs. The responsivity of a PD is the ratio of the output photocurrent to the incident light power. It can be related to EQE as [23]:

$$R = \text{EQE}\frac{q\lambda}{hc} \tag{1}$$

The FWHM is an important figure of merit for the narrowband PD as it defines the specificity of the detection wavelength as well as the imaging resolution.

3.1. Impact of Trap Density and Thickness of Perovskite Layer

First, we test the impact of trap density (N_t) and thickness (d) of the perovskite film as the carrier lifetime, and thickness of the perovskite layer have to be designed carefully. This step is essential to check and ensure the design criterion of suppression of short wavelength range. In these simulations, the energy gap of the perovskite is set at 1.62 eV. All other parameters are fixed as in Tables 1 and 2 unless otherwise stated. In this regard, the simulated external quantum efficiency (EQE) curves under 0 V bias are shown for different N_t and d values in Figure 3a and 3b, respectively. Upon decreasing N_t, the carrier lifetime (τ) increases. On the other hand, as long as d is constant, the transition time (t_{tr}) is constant, which is given by,

$$t_{tr} = \frac{d^2}{\mu V} \tag{2}$$

where μ is the carrier mobility and V is the average electric potential along the perovskite layer. The calculated value of t_{tr} is about 25 ns at d = 2.5 μm as shown in Figure 3a. When the transition time is much longer than the carrier lifetime, the photogenerated carriers recombine before collection occurs. This can be seen for higher values of N_t. However, when N_t is declined (down to 1×10^{13} cm^{-3}), the transition time becomes lower than the carrier lifetime implying lower recombination probability, which results in higher EQE [23]. That is why it is crucial to keep high defect density in the perovskite film for an appropriate narrowband PD design. Moreover, the thickness of perovskite film is varied while keeping a fixed N_t = 1×10^{15} cm^{-3} (and so a fixed τ = 1 ns). The results show that when decreasing d, the transition time will drop as shown in Figure 3b. Once the transition time becomes comparable to the carrier lifetime, the carrier recombination likelihood decreases before the collection resulting in an increased EQE at the short wavelength. So, a thickness of perovskite film larger than 2 μm could be suitable to achieve an adequate narrowband PD operation.

To present more physical insight into the photo response of the PD in accordance with wavelength variation, we draw the profiles of the normalized photon absorption rate in two different cases as shown in Figure 4. The first case is for a perovskite thickness d = 2.5 μm, while the other one is for a thinner perovskite layer for which d = 0.5 μm. For the two cases, N_t = 1×10^{15} cm^{-3}. As it can be inferred from Figure 4a, the distribution of photon absorption reveals that the part of incident light with wavelengths range shorter than about 700 nm can be completely absorbed primarily by the perovskite film. This results in the dominance of generation and recombination inside the perovskite and translates into low EQE below 700 nm, as indicated in Figure 3a. The fringes that originate from the interference progressively become noticeable for wavelengths longer than 700 nm up to about 850 nm. This means that the generation and recombination occur mainly inside the organic film, implying less absorption in the perovskite layer at higher wavelengths, which results in higher EQE. On the other hand, for thin perovskite film, the profile of the normalized photon absorption, displayed in Figure 4b, reveals that there is a penetration of photons in the organic layer for the wavelength range from 500 nm up to 700 nm. This result is supported by the behavior of the EQE shown in Figure 3b.

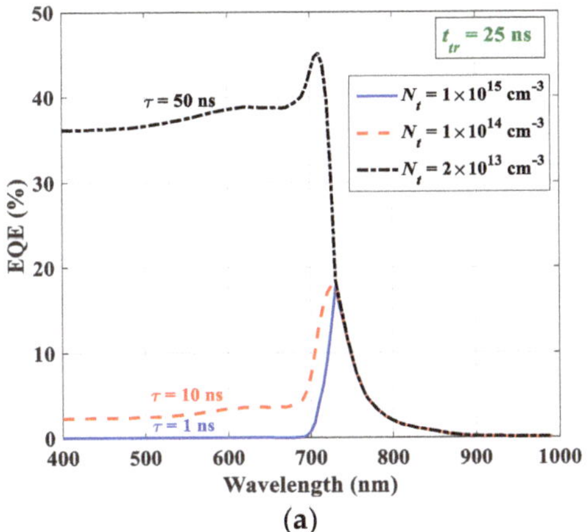

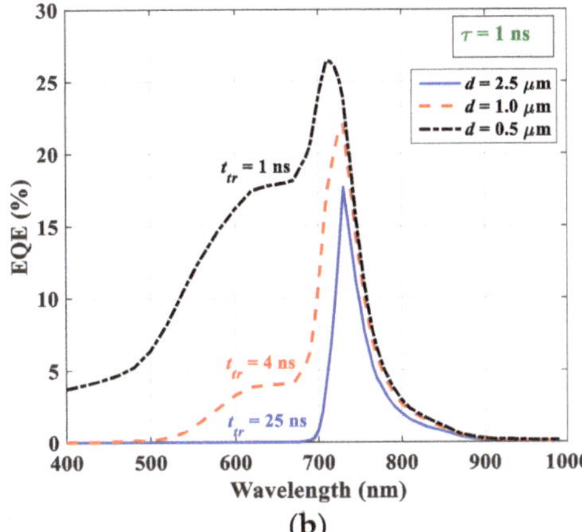

Figure 3. EQE Simulation results for different values of (**a**) trap densities and (**b**) thickness of perovskite layer.

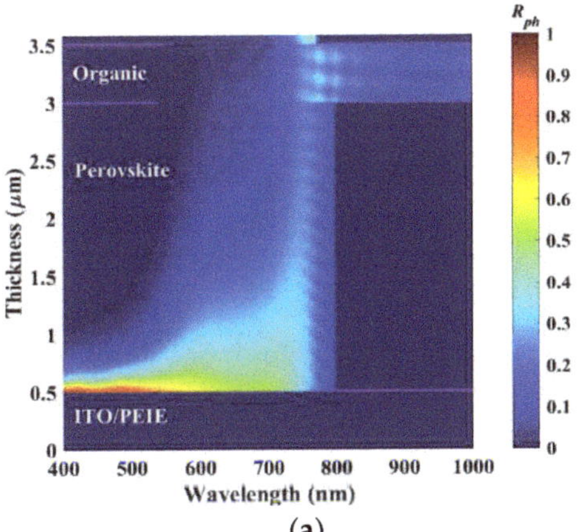

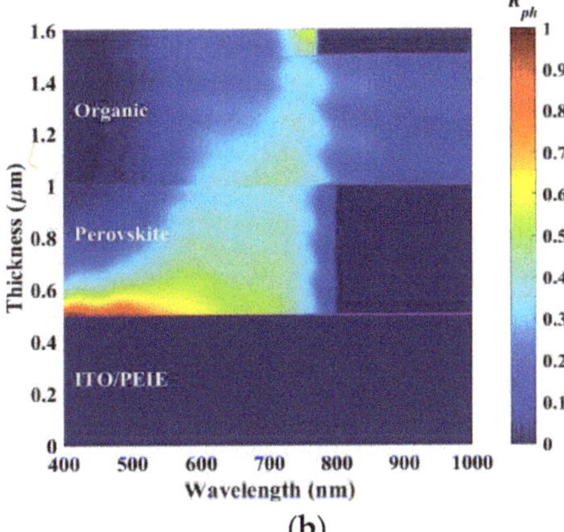

Figure 4. Simulated normalized nonlinear contours of photon absorption rate with a perovskite thickness of (**a**) 2.5 μm and (**b**) 0.5 μm.

3.2. Impact of Energy Gap of Perovskite Material

As mentioned herein, $Cs_yFA_{1-y}Pb(I_xBr_{1-x})_3$ perovskite material is selected thanks to its bandgap tunability. In the following analysis, the impact of the energy gap of $Cs_yFA_{1-y}Pb(I_xBr_{1-x})_3$ on the photoresponsivity is examined. Figure 5a shows the EQE curves for different values of E_g. As E_g increases, the EQE increases as well; however, the FWHM increases, which means a broader band. For a quantitative difference between the various bandgap cases, we draw the peak responsivity and FWHM as seen in Figure 5b. It can be observed that the peak responsivity increases upon increasing E_g while there is an optimum minimum value of 33 nm that occurs at E_g = 1.69 eV. So, a suitable design can be chosen for which the energy gap of the perovskite layer is 1.69 eV. For this case, the peak responsivity is 0.12 A/W at a full width at half maximum of 33 nm.

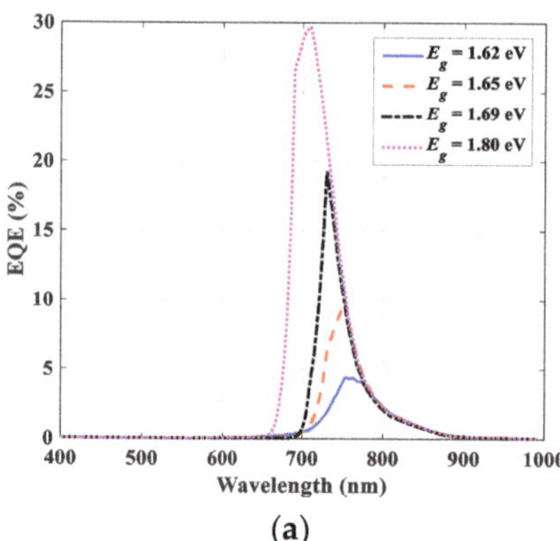

(a)

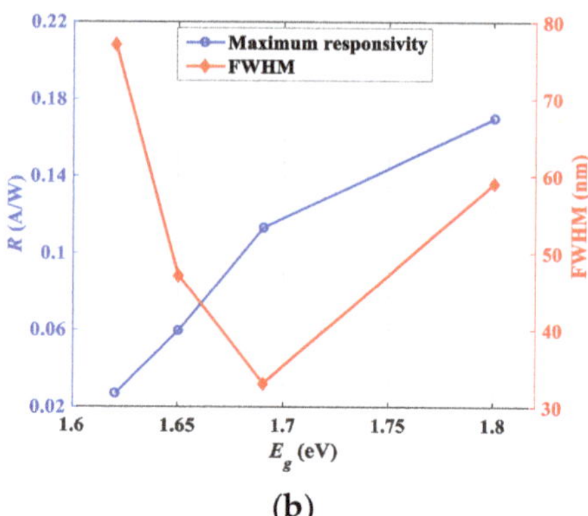

(b)

Figure 5. Impact of perovskite energy gap on (**a**) the EQE characteristics, and (**b**) peak responsivity and full width at half maximum.

3.3. Impact of Reverse Bias

In this subsection, the influence of reverse bias on the PD performance is presented. In these simulations, the perovskite energy gap is taken to be 1.69 eV. Figure 6a displays the EQE for three cases of reverse bias at 0, 2, and 5 V. As the reverse bias rises, the EQE increases but the spectrum becomes wider. For a quantitative view on the biasing conditions, the peak responsivity and FWHM are plotted as shown in Figure 6b. As the reverse bias increases, the two PD parameters raise; however, the rate of rise is slowed. A suitable design could be achieved for a powered PD at a reverse voltage of 5 V. In this case, the peak responsivity and the FWHM are about 0.34 A/W and 41 nm, respectively.

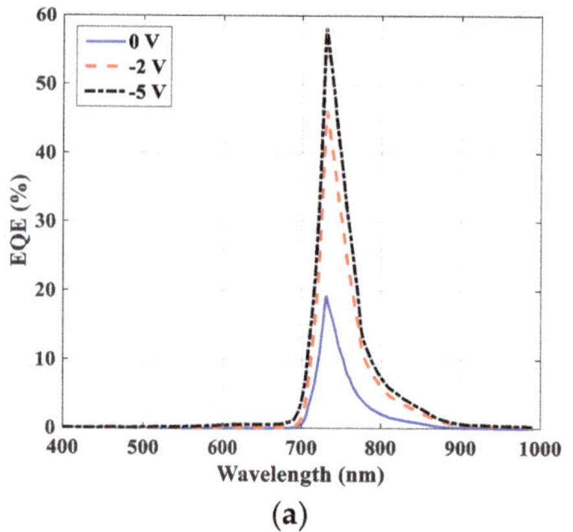

(a)

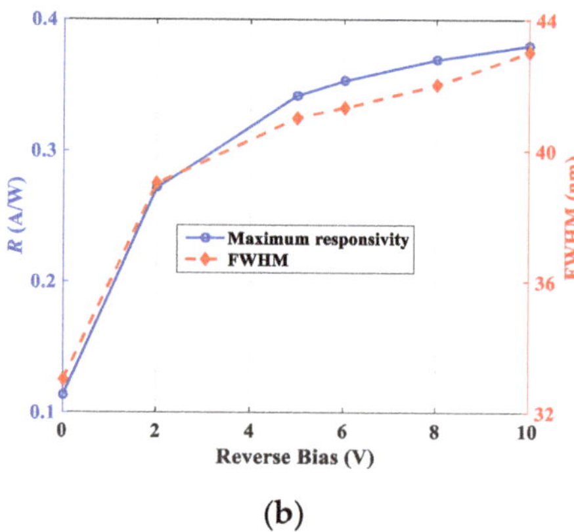

(b)

Figure 6. Impact of reverse bias on (**a**) the EQE characteristics and (**b**) peak responsivity and full width at half maximum.

3.4. Impact of Organic Layer Thickness and Defects

Here, we provide a parametric study to examine the variation of the thickness and trap density of the organic layer on the responsivity and FWHM. The thickness is varied from 100 to 500 nm while the trap density is weighted by a factor from 5% to 100%. As seen in Figure 7, when the trap density declines, the responsivity increases (see Figure 7a); however, the FWHM trend is different (see Figure 7b). The FWHM is maximum for lower values of thickness given a fixed trap density. It is obvious that if the design thickness is at 500 nm, an optimum choice could be met for lower trap density. For a selected case of 10%

reduction of the trap densities listed in Table 2, the responsivity is 0.3 A/W and the FWHM is 37 nm. Although the FWHM has increased by about 12% of the corresponding value at zero bias, an enhancement of 170% in R is achieved. This result shows the importance of decreasing the bulk defects in organic layers. With the continuous advancement in processing techniques, organic semiconductors can be enabled to have lower trap density and improved crystallinity.

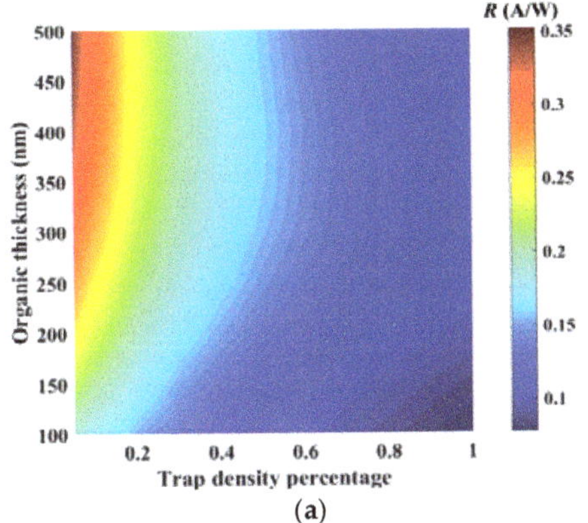

(a)

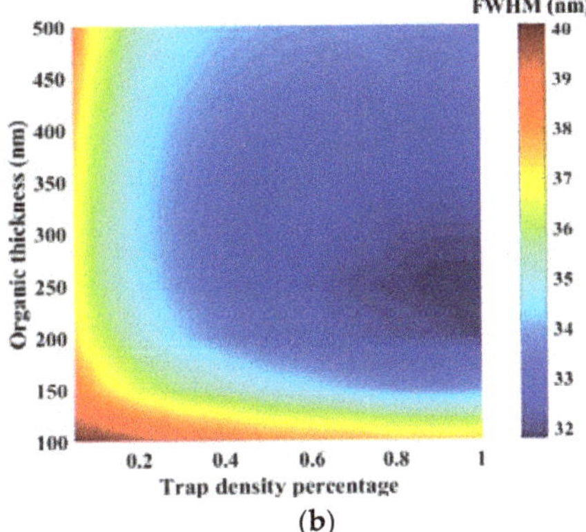

(b)

Figure 7. Contour plots representing the impact of organic thickness and trap density on (**a**) peak responsivity and (**b**) FWHM.

Finally, a comparison between the key detector parameters of our structure and those of some organic and inorganic narrowband PDs is provided in Table 3. The PDs are divided into groups where the first group regards organic-base PDs showing structures of narrow band [40,41] and some others with a higher detection band [42,43]. Although the bias increases R and EQE, it results in widening the band [43]. The second category considers the hybrid perovskite/organic based PDs showing lower responsivity in comparison to organic candidates [21,22]. This is due to the thick perovskite layer, which generally reduces the photogeneration inside the organic film. In our hybrid system, the simulation results are encouraging as the value of R is comparable to organic PD while the FWHM is very narrow. The influence of bias is to increase R and FWHM besides it causes a redshift in the detection peak wavelength (λ_{peak}). Moreover, reducing organic bulk defects results in a substantial rise in R, while the FWHM does not significantly degrade as evident from the results listed in Table 3.

Table 3. State-of-the-art comparison showing the main metrics of some reported narrow-band NIR organic-based PDs.

Category	Active Materials	λ_{Peak} (nm)	FWHM (nm)	EQE (%)	R (A/W)	Bias (V)	REF
Organic based	PBTTT:PC_{61}BM	775	15	40.0	0.250	0	[40]
	PCDTPTSe:PC_{71}BM	710	60	18.0	0.100	0	[41]
	PolyTPD:SBDTIC	740	141	10.5	0.060	0	[42]
	PCbisDPP:PC61BM	730	210	80.0	0.310	−3.0	[43]
Hybrid Perovskite/Organic	$CH_3NH_3PbI_3$ /PCPDTBT:PC_{71}BM	830	98	4.20	0.027	0	[22]
	CH3NH3Pb$I_{3\text{-}x}Br_x$/PTB7	690	50	20.0	0.110	−0.1	[21]
Our hybrid Perovskite/Organic	$Cs_yFA_{1-y}Pb(I_xBr_{1-x})_3$/Blend	730	33	19.2	0.113	0	
	$Cs_yFA_{1-y}Pb(I_xBr_{1-x})_3$/Blend	707	41	58.0	0.340	−5	
	$Cs_yFA_{1-y}Pb(I_xBr_{1-x})_3$/Blend (Low organic defects)	730	37	35.5	0.303	0	

Regarding inorganic PD candidates, a recent research study revealed a responsivity is 0.09 A/W with a FWHM of 35.3 nm at a peak wavelength of 735 nm for the $CdSe/Sb_2(S_{1-x},Se_x)_3$ inorganic system [24]. Thus, compared to our results, the proposed perovskite/organic PD is promising and, upon appropriate design and simulation-driven experimental studies, could compete the inorganic system PDs.

4. Conclusions

In this paper, we report on the results of our simulation studies on the perovskite/organic heterojunction photodetector. A model structure of perovskite/organic heterojunction PD was built in Silvaco TCAD environment. The effects of perovskite layer thickness, trap density, and energy gap on the performance of the proposed PD are discussed. The simulations are conducted regarding the external quantum efficiency (EQE) curves. Besides, the main performance parameters like responsivity and full width at half maximum are also presented. The simulation results show that the proposed PD reveals optimum performance when the thickness of perovskite and organic layers are 2.5 μm and 500 nm, respectively. The FWHM has a minimum value of 33 nm at an energy gap of 1.69 eV of the perovskite material. The effect of reverse bias is also demonstrated, showing that a proper design could be accomplished for a powered PD at a reverse voltage of 5 V. Under this condition, the peak responsivity and the FWHM are about 0.34 A/W and 41 nm, respectively. Reducing the bulk trap density inside the organic has a positive effect, as R is substantially increased while the FWHM degrades by only 12%. This TCAD simulation study shows promising results and reveals that the design of perovskite/organic PD could be feasible. Additionally, the design concepts, presented in this paper, could be easily extended to other perovskite and organic partner materials.

Supplementary Materials: The following supporting information can be downloaded at: https://www.mdpi.com/article/10.3390/cryst12081033/s1, Figure S1: Optical constants (n and k) of (a) ITO and (b) PEDOT:PSS; Figure S2: Optical constants (n and k) of organic blend PBDTTT-c:C60-PCBM; Figure S3: Optical constants (n and k) of $Cs_yFA_{1-y}Pb(I_xBr_{1-x})_3$ by Ellipsometric measurements for (a) E_g = 1.62 eV, (b) E_g = 1.65, (c) E_g = 1.69 eV, and (d) E_g = 1.8 eV; Figure S4: Silvaco Script used to calibrate the wideband organic PD.

Author Contributions: Conceptualization, M.S.S., A.S. and M.T.A.; formal analysis, A.H.A.-B., G.M.A. and M.S.O.; software, M.S.S., A.S. and M.T.A.; investigation, M.S.S., A.S. and M.S.O.; methodology, M.S.S., M.T.A. and M.F.; supervision, A.S., A.H.A.-B. and M.T.A.; validation, M.S.S., G.M.A., A.H.A.-B. and M.F.; writing—original draft, M.S.S., A.H.A.-B. and G.M.A.; writing—review and editing, M.S.S., A.S. and M.S.O. All authors have read and agreed to the published version of the manuscript.

Funding: This research has been funded by Scientific Research Deanship at University of Ha'il–Saudi Arabia through project number RG-21 094.

Data Availability Statement: Not applicable.

Acknowledgments: The authors would like to express their great appreciation to Julie Euvrard for providing the experimental optical constants of the organic blend material used in our study.

Conflicts of Interest: The authors declare no conflict of interest.

References

1. Wang, Y.; Kublitski, J.; Xing, S.; Dollinger, F.; Spoltore, D.; Benduhn, J.; Leo, K. Narrowband organic photodetectors–towards miniaturized, spectroscopic sensing. *Mater. Horiz.* **2022**, *9*, 220–251. [CrossRef] [PubMed]
2. Wu, Y.L.; Fukuda, K.; Yokota, T.; Someya, T. A highly responsive organic image sensor based on a two-terminal organic photodetector with photomultiplication. *Adv. Mater.* **2019**, *31*, 1903687. [CrossRef] [PubMed]
3. Han, Y.; Wen, R.; Zhang, F.; Shi, L.; Wang, W.; Ji, T.; Cui, Y. Photodetector Based on CsPbBr3/Cs4PbBr6 Composite Nanocrystals with High Detectivity. *Crystals* **2021**, *11*, 1287. [CrossRef]
4. Wang, Y.; Benduhn, J.; Baisinger, L.; Lungenschmied, C.; Leo, K.; Spoltore, D. Optical distance measurement based on induced nonlinear photoresponse of high-performance organic near-infrared photodetectors. *ACS Appl. Mater. Interfaces* **2021**, *13*, 23239–23246. [CrossRef]

5. Babics, M.; Bristow, H.; Zhang, W.; Wadsworth, A.; Neophytou, M.; Gasparini, N.; McCulloch, I. Non-fullerene-based organic photodetectors for infrared communication. *J. Mater. Chem. C* **2021**, *9*, 2375–2380. [CrossRef]
6. Ismail, R.A.; Abdulnabi, R.K.; Abdulrazzaq, O.A.; Jawad, M.F. Preparation of MAPbI3 perovskite film by pulsed laser deposition for high-performance silicon-based heterojunction photodetector. *Opt. Mater.* **2022**, *126*, 112147. [CrossRef]
7. Kim, J.J.; Wang, Y.; Wang, H.; Lee, S.; Yokota, T.; Someya, T. Skin electronics: Next-generation device platform for virtual and augmented reality. *Adv. Funct. Mater.* **2021**, *31*, 2009602. [CrossRef]
8. Kabir, S.; Takayashiki, Y.; Ohno, A.; Hanna, J.I.; Iino, H. Near-infrared organic photodetectors with a soluble Alkoxy-Phthalocyanine derivative. *Opt. Mater.* **2022**, *31*, 112209. [CrossRef]
9. Nikovskiy, I.A.; Isakovskaya, K.L.; Nelyubina, Y.V. New Low-Dimensional Hybrid Perovskitoids Based on Lead Bromide with Organic Cations from Charge-Transfer Complexes. *Crystals* **2021**, *11*, 1424. [CrossRef]
10. Liu, M.; Wang, J.; Zhao, Z.; Yang, K.; Durand, P.; Ceugniet, F.; Ulrich, G.; Niu, L.; Ma, Y.; Leclerc, N.; et al. Ultra-Narrow-Band NIR Photomultiplication Organic Photodetectors Based on Charge Injection Narrowing. *J. Phys. Chem. Lett.* **2021**, *12*, 2937–2943. [CrossRef]
11. Liu, M.Y.; Wang, J.; Yang, K.X.; Liu, M.; Zhao, Z.J.; Zhang, F.J. Broadband photomultiplication organic photodetectors. *Phys. Chem. Chem. Phys.* **2021**, *23*, 2923–2929. [CrossRef] [PubMed]
12. Weng, S.; Zhao, M.; Jiang, D. Organic Ternary Bulk Heterojunction Broadband Photodetectors Based on Nonfullerene Acceptors with a Spectral Response Range from 200 to 1050 nm. *J. Phys. Chem. C* **2021**, *125*, 20676–20685. [CrossRef]
13. Yu, Y.Y.; Peng, Y.C.; Chiu, Y.C.; Liu, S.J.; Chen, C.P. Realizing Broadband NIR Photodetection and Ultrahigh Responsivity with Ternary Blend Organic Photodetector. *Nanomaterials* **2022**, *12*, 1378. [CrossRef] [PubMed]
14. Zhao, Z.; Xu, C.; Niu, L.; Zhang, X.; Zhang, F. Recent progress on broadband organic photodetectors and their applications. *Laser Photonics Rev.* **2020**, *14*, 2000262. [CrossRef]
15. Kim, J.; So, C.; Kang, M.; Sim, K.M.; Lim, B.; Chung, D.S. A regioregular donor–acceptor copolymer allowing a high gain–bandwidth product to be obtained in photomultiplication-type organic photodiodes. *Mater. Horiz.* **2021**, *8*, 276–283. [CrossRef]
16. Shan, C.; Meng, F.; Yu, J.; Wang, Z.; Li, W.; Fan, D.; Kyaw, A.K.K. An ultrafast-response and high-detectivity self-powered perovskite photodetector based on a triazine-derived star-shaped small molecule as a dopant-free hole transporting layer. *J. Mater. Chem. C* **2021**, *9*, 7632–7642. [CrossRef]
17. Salah, M.M.; Abouelatta, M.; Shaker, A.; Hassan, K.M.; Saeed, A. A comprehensive simulation study of hybrid halide perovskite solar cell with copper oxide as HTM. *Semicond. Sci. Technol.* **2019**, *34*, 115009. [CrossRef]
18. Basyoni, M.S.S.; Salah, M.M.; Mousa, M.; Shaker, A.; Zekry, A.; Abouelatta, M.; Gontrand, C. On the Investigation of Interface Defects of Solar Cells: Lead-Based vs Lead-Free Perovskite. *IEEE Access* **2021**, *9*, 130221–130232. [CrossRef]
19. Wu, G.; Fu, R.; Chen, J.; Yang, W.; Ren, J.; Guo, X.; Chen, H. Perovskite/Organic Bulk-Heterojunction Integrated Ultrasensitive Broadband Photodetectors with High Near-Infrared External Quantum Efficiency over 70%. *Small* **2018**, *14*, 1802349. [CrossRef]
20. Sun, H.; Tian, W.; Cao, F.; Xiong, J.; Li, L. Ultrahigh-performance self-powered flexible double-twisted fibrous broadband perovskite photodetector. *Adv. Mater.* **2018**, *30*, 1706986. [CrossRef]
21. Qin, Z.; Song, D.; Xu, Z.; Qiao, B.; Huang, D.; Zhao, S. Filterless narrowband photodetectors employing perovskite/polymer synergetic layers with tunable spectral response. *Org. Electron.* **2020**, *76*, 105417. [CrossRef]
22. Lan, Z.; Cai, L.; Luo, D.; Zhu, F. Narrowband near infrared perovskite/polymer hybrid photodetectors. *ACS Appl. Mater. Interfaces* **2020**, *13*, 981–988. [CrossRef] [PubMed]
23. Cao, F.; Chen, J.; Yu, D.; Wang, S.; Xu, X.; Liu, J.; Zeng, H. Bionic detectors based on low-bandgap inorganic perovskite for selective NIR-I photon detection and imaging. *Adv. Mater.* **2020**, *32*, 1905362. [CrossRef] [PubMed]
24. Li, K.; Lu, Y.; Yang, X.; Fu, L.; He, J.; Lin, X.; Tang, J. Filter-free self-power CdSe/Sb2(S1−x,Sex)3 nearinfrared narrowband detection and imaging. *InfoMat* **2021**, *3*, 1145–1153. [CrossRef]
25. Wang, L.; Li, Z.; Li, M.; Li, S.; Lu, Y.; Qi, N.; Luo, L.B. Self-powered filterless narrow-band p–n heterojunction photodetector for low background limited near-infrared image sensor application. *ACS Appl. Mater. Interfaces* **2020**, *12*, 21845–21853. [CrossRef] [PubMed]
26. Scheer, R.; Schock, H.W. *Chalcogenide Photovoltaics: Physics, Technologies, and Thin Film Devices*; John Wiley & Sons: Berlin, Germany, 2011.
27. Atlas User's Manual, Silvaco Inc., Santa Clara. USA. Available online: https://silvaco.com/products/tcad/device_simulation/atlas/atlas.html (accessed on 1 June 2022).
28. Werner, J.; Nogay, G.; Sahli, F.; Yang, T.C.J. Complex refractive indices of cesium–formamidinium-based mixed-halide perovskites with optical band gaps from 1.5 to 1.8 eV. *ACS Energy Lett.* **2018**, *3*, 742–747. [CrossRef]
29. Euvrard, J.; Revaux, A.; Cantarano, A.; Jacob, S.; Kahn, A.; Vuillaume, D. Impact of unintentional oxygen doping on organic photodetectors. *Org. Electron.* **2018**, *54*, 64–71. [CrossRef]
30. Bandelow, U. *Optoelectronic Devices—Advanced Simulation and Analysis*; Piprek, J., Ed.; Springer: New York, NY, USA, 2005; pp. 63–85.
31. Yu, H.; Ji, S.; Luo, X.; Xie, Q. Technology CAD (TCAD) Simulations of Mg2Si/Si Heterojunction Photodetector Based on the Thickness Effect. *Sensors* **2021**, *21*, 5559. [CrossRef]
32. Yu, H.; Gao, C.; Zou, J.; Yang, W.; Xie, Q. Simulation Study on the Effect of Doping Concentrations on the Photodetection Properties of Mg2Si/Si Heterojunction Photodetector. *Photonics* **2021**, *8*, 509. [CrossRef]

33. Beal, R.E.; Slotcavage, D.J.; Leijtens, T.; Bowring, A.R.; Belisle, R.A.; Nguyen, W.H.; McGehee, M.D. Cesium lead halide perovskites with improved stability for tandem solar cells. *J. Phys. Chem. Lett.* **2016**, *7*, 746–751. [CrossRef]
34. Abdelaziz, W.; Zekry, A.; Shaker, A.; Abouelatta, M. Numerical study of organic graded bulk heterojunction solar cell using SCAPS simulation. *Sol. Energy* **2020**, *211*, 375–382. [CrossRef]
35. Gamal, N.; Sedky, S.H.; Shaker, A.; Fedawy, M. Design of lead-free perovskite solar cell using Zn1-xMgxO as ETL: SCAPS device simulation. *Optik* **2021**, *242*, 167306. [CrossRef]
36. Jafarzadeh, F.; Aghili, H.; Nikbakht, H.; Javadpour, S. Design and optimization of highly efficient perovskite/homojunction SnS tandem solar cells using SCAPS-1D. *Sol. Energy* **2022**, *236*, 195–205. [CrossRef]
37. Khadka, D.B.; Shirai, Y.; Yanagida, M.; Ryan, J.W.; Miyano, K. Exploring the effects of interfacial carrier transport layers on device performance and optoelectronic properties of planar perovskite solar cells. *J. Mater. Chem. C* **2017**, *5*, 8819–8827. [CrossRef]
38. Lienhard, P. Caractérisation et modélisation du vieillissement des Photodiodes Organiques. Ph.D. Thesis, Université Grenoble Alpes, Grenoble, France, 2016.
39. Herrbach-Euvrard, J. Organic Semiconductor P-Doping: Toward a Better Understanding of the Doping Mechanisms and Integration of the P-Doped Layer in Organic Photodetectors. Ph.D. Thesis, Lille, France, 2017. Available online: https://www.semanticscholar.org/paper/Organic-semiconductor-p-doping-%3A-toward-a-better-of-Herrbach-Euvrard/526f08583a17ee82906336c5b9fab0c1972ac29b (accessed on 30 June 2022).
40. Tang, Z.; Ma, Z.; Sánchez-Díaz, A.; Ullbrich, S.; Liu, Y.; Siegmund, B.; Vandewal, K. Polymer: Fullerene bimolecular crystals for near-infrared spectroscopic photodetectors. *Adv. Mater.* **2017**, *29*, 1702184. [CrossRef] [PubMed]
41. Yang, J.; Huang, J.; Li, R.; Li, H.; Sun, B.; Lin, Q.; Tang, Z. Cavity-Enhanced Near-Infrared Organic Photodetectors Based on a Conjugated Polymer Containing [1, 2, 5] Selenadiazolo [3, 4-c] Pyridine. *Chem. Mater.* **2021**, *33*, 5147–5155. [CrossRef]
42. Xia, K.; Li, Y.; Wang, Y.; Portilla, L.; Pecunia, V. Narrowband-Absorption-Type Organic Photodetectors for the Far-Red Range Based on Fullerene-Free Bulk Heterojunctions. *Adv. Opt. Mater.* **2020**, *8*, 1902056. [CrossRef]
43. Xia Opoku, H.; Lim, B.; Shin, E.S.; Kong, H.; Park, J.M.; Bathula, C.; Noh, Y.Y. Bis-Diketopyrrolopyrrole and Carbazole-Based Terpolymer for High Performance Organic Field-Effect Transistors and Infra-Red Photodiodes. *Macromol. Chem. Phys.* **2019**, *220*, 1900287. [CrossRef]

 *crystals* 

Article

External Electric Field Tailored Spatial Coherence of Random Lasing

Yaoxing Bian [1,2], Hongyu Yuan [1], Junying Zhao [1], Dahe Liu [1], Wenping Gong [1,*] and Zhaona Wang [1,*]

1 Applied Optics Beijing Area Major Laboratory, Department of Physics, Beijing Normal University, Beijing 100875, China
2 College of Physics and Optoelectronics, Taiyuan University of Technology, Taiyuan 030024, China
* Correspondence: wpgong@bnu.edu.cn (W.G.); zhnwang@bnu.edu.cn (Z.W.)

Abstract: In this study, spatial coherence tunable random lasing is proposed by designing a random laser with separate coupling configuration between the gain medium and the scattering part. By using the polymer dispersion liquid crystal (PDLC) film with tunable scattering coefficient for supplying random scattering feedback and output modification, red, green and blue random lasers are obtained. By applying or removing electric field to manipulate the scattering intensity of the PDLC film, intensity and spatial coherence of these random lasing are then switched between the high or low state. This work demonstrates that controlling the external scattering intensity is an effective method to manipulate the spatial coherence of random lasing.

Keywords: random laser; spatial coherence; random scattering feedback; tunable scattering coefficient

Citation: Bian, Y.; Yuan, H.; Zhao, J.; Liu, D.; Gong, W.; Wang, Z. External Electric Field Tailored Spatial Coherence of Random Lasing. *Crystals* **2022**, *12*, 1160. https://doi.org/10.3390/cryst12081160

Academic Editor: Shin-Tson Wu

Received: 18 June 2022
Accepted: 4 August 2022
Published: 18 August 2022

Publisher's Note: MDPI stays neutral with regard to jurisdictional claims in published maps and institutional affiliations.

1. Introduction

As an unconventional laser, random lasing with low spatial coherence and high brightness can effectively suppress speckle noise [1], and plays a vital role in the fields of optical imaging [2–4], display [5], sensing [6,7], information security [8,9] and integrated photoelectron [10]. The optical field of random lasing is endowed with rich physical characteristics due to the inherent randomness of the multiple-scattering feedback mechanism [11–13]. Therefore, the optical field manipulation of random lasing has become an important research topic [14–16]. In particular, low spatial coherence is the unique characteristics of random lasing different from the high spatial coherence of traditional lasing. Regulating the spatial coherence of random lasing is conducive to broaden its application fields [17,18]. At present, researchers have regulated the spatial coherence of random lasing by changing the scattering mean free path [19], pump area [20] and pump power density [21]. The method of controlling the scattering mean free path is widely used for tuning random lasing by changing the scatterer concentration in the gain medium due to its simple operation [19]. However, this manipulation method is not conducive to dynamically regulating the spatial coherence of random lasing because of the configuration of directly mixing the gain medium and the scattering material together. Recently, the separate coupling of individual gain medium and individual scattering part has been reported as a new configuration for flexibly regulating the random lasing pulses in the time domain [22], demonstrating a huge superiority in dynamic regulation. As a new coupling configuration, the effect of external scattering intensity on the random lasing has not been reported. The potential of this separate coupling configuration in dynamically regulating the spatial coherence of random lasing needs to be further revealed. Therefore, it is necessary to propose a random laser with a separate coupling configuration to realize spatial coherence tunable random lasing through dynamically adjusting the scattering coefficient of the scattering part.

In this work, a random laser with dynamically tunable spatial coherence is proposed based on the separate coupling between gain medium and scattering part. The PDLC film with adjustable scattering coefficient is coupled with different dye solutions for colorful random lasing. The coherent feedback random lasing with wavelengths of 457.5 nm, 542.7 nm and 662.9 nm are realized. Furthermore, the effect of the scattering coefficient on the optical field of random lasing has also been studied. A small scattering coefficient means a high intensity random lasing with high spatial coherence. Therefore, the intensity and spatial coherence of random lasing can be flexibly manipulated by applying or removing the electric field on the PDLC film. This work can further expand the application of random lasing in dynamic imaging, display and other fields.

2. Materials and Methods

The gain dyes selected in the experiments were Coumarin 1(C1, ACROS), Coumarin 153 (C153, Tokyi Chemical Industry, Tokyo, Japan) and 4-(dicyanomethylene)-2-tert-butyl-6-(1,1,7,7-tetramethyljulolidin-4-yl-vinyl)-4H-pyran (DCJTB, Tokyi Chemical Industry, Tokyo, Japan). Firstly, these gain materials were diluted in ethanol to obtain the C1 solution with a concentrate of 1.0 mg mL^{-1}, the C153 solution at 1.5 mg mL^{-1} and the DCJTB solution at 1.5 mg mL^{-1} for further operation. Then, 1.50 mL dye solution was added into the quartz cuvette as the gain part. Finally, the commercial PDLC film with a thickness of 400 μm was coupled to the outside of the quartz cuvette as the tunable scattering part of the random laser, and the average size of LC droplets was about 400 nm.

The schematic diagram and optical photo of the experimental setup are shown in Figure 1a,b, respectively. The samples were optically excited by a frequency-doubled and Q-switched neodymium doped yttrium aluminum garnet (Nd:YAG) laser with a wavelength of 355 nm, repetition rate of 10 Hz, and pulse duration of 8 ns. The pump power density was controlled by an adjustable attenuation element. The dye solution was pumped vertically from the top of the quartz cuvette by the pulses with a spot diameter of 8.0 mm. The scattering coefficient of the PDLC film was regulated by applying and removing the AC voltage of 40 V on the PDLC film. The spectra of emission light passing through PDLC film were measured by a spectrometer of Ocean Optics model Maya Pro 2000 with a spectral resolution of 0.4 nm and an integration time of 100 ms. All experiments were performed at room temperature.

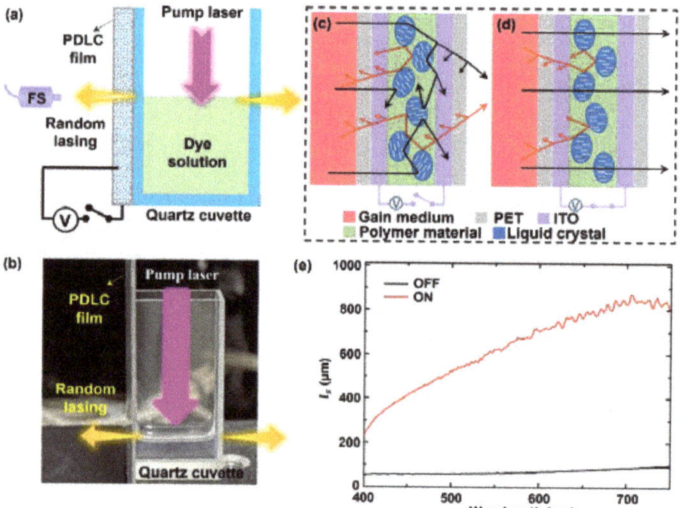

Figure 1. Schematic diagram of spatial coherence-tunable random lasing. (**a**,**b**) The schematic diagram (**a**) and the optical photo (**b**) of the experimental setup. (**c**,**d**) The emission principle diagram of random lasing in the coupling system with the gain medium and the PDLC film without (**c**) and with (**d**) the external electric field. (**e**) The calculated scattering mean free path (l_s) when the electric field was switched on (ON) and switched off (OFF), respectively.

3. Results and Discussion

PDLC is widely applied in optical devices due to the excellent tuning possibilities and extreme form of anisotropic scattering [23]. PDLC film is a composite film made by mixing nematic liquid crystal (LC) and polymer in a certain proportion, which is sandwiched between two transparent conductive films with Polyethylene terephthalate (PET) and Indium tin oxide (ITO). The refractive index difference between PET and ITO films and between ITO and air enables them to reflect the incident light. Here, the used LC was an anisotropic material with refractive index n_o (ordinary) and n_e (extraordinary) for further tuning the scattering coefficient of the PDLC film through external bias voltage. The mixed polymer with refractive index n_p ($n_o \approx n_p$) was used to form the PDLC film. Approximately, for a LC droplet with an orientation angle θ of LC molecule relative to the light polarization direction, the equivalent refractive index n_{eff} can be expressed as $\frac{1}{n_{eff}^2} = \frac{\cos^2\theta}{n_o^2} + \frac{\sin^2\theta}{n_e^2}$ [24,25]. The mismatch of refraction index between the LC droplet and polymer can be written as $\Delta n = n_{eff} - n_o$, which can be controlled by the LC molecule orientation. Therefore, the scattering coefficient of PDLC film can be controlled through using the electric field to change the LC orientation and then induce the mismatch of refractive index between polymer and LC.

The design principle of the spatial coherence tunable random lasing is schematically presented in Figure 1c,d by coupling the mentioned PDLC film and gain region with varying dye solution. When the gain medium is pumped, the emission photons are radiated from the gain region. Some of them are reflected and scattered to the gain region by the PDLC film, supplying random feedback for the generation of random lasing. Some of the emission photons pass through the PDLC film, superimposing additional modification in random phase and intensity. The emitted random lasing is thus obtained and its performance can be modified by the scattering property of the PDLC film. When the PDLC film is without electric field (Figure 1c, OFF), the orientation of LC molecules is highly random, inducing strong scattering due to the highly mismatched refractive index between polymer and LC droplet. Thus, the PDLC film presents an opaque state, demonstrating a small scattering mean free path (l_s) obtained through measuring the extinction coefficient [26] as shown in Figure 1e. This PDLC film in OFF state supplies a strong backscattering feedback for the gain part and strong tuning capability of emission light in spatial coherence. Thus, the emitted random lasing has low intensity and spatial coherence. When the external electric field is applied to the PDLC film (Figure 1d, ON), the orientation of LC molecules is almost consistent with the direction of electric field. The corresponding PDLC film is transparent and has a relatively large scattering mean free path (Figure 1e). The weak backscattering feedback is for the gain part and a slight decrease acts on the spatial coherence of the forward emission light. As a result, the emitted random lasing has high intensity and spatial coherence. Therefore, the spatial coherence of the proposed random lasing can be flexibly regulated by switching ON/OFF the applied electric field.

The effect of scattering feedback of individual scattering part on random lasing is demonstrated by measuring the emission spectra of the random lasers under different pump power densities in Figure 2. A blue random laser was obtained by coupling the C1 solution and PDLC film, and named as RL_B. The emission spectra of RL_B at different pump power densities when the external electric field is removed from the PDLC film in OFF state are shown in Figure 2a. Only broadband fluorescence spectra can be observed as the pump power density is less than 0.30 MW cm^{-2}. However, when the pump power density is greater than 0.30 MW cm^{-2}, many narrow linewidth peaks can be observed around 457.5 nm. The results show that the emission light from the C1 solution is randomly scattered by the PDLC film and the coherent resonance is established in the coupling system. Meanwhile, the variations of integral intensity of the emission spectra with pump power density are shown in Figure 2c. When the pump power density is greater than 0.30 MW cm^{-2}, the integral intensity increases sharply with increasing pump power density, which indicates that the threshold of the random lasing is about 0.30 MW cm^{-2} [27,28].

The emission spectra of RL_B at different pump power densities when the electric field is applied to the PDLC film in ON state are shown in Figure 2b. Coherent feedback random lasing can also be obtained when the pump power density is greater than 0.20 MW cm^{-2}, and the intensity is much higher than that of PDLC film in OFF state. The corresponding threshold of RL_B is about 0.20 MW cm^{-2} (Figure 2c), which is smaller than that of the PDLC film in OFF state. The increase in intensity and decrease in threshold are due to the weak scattering feedback of the PDLC film in ON state.

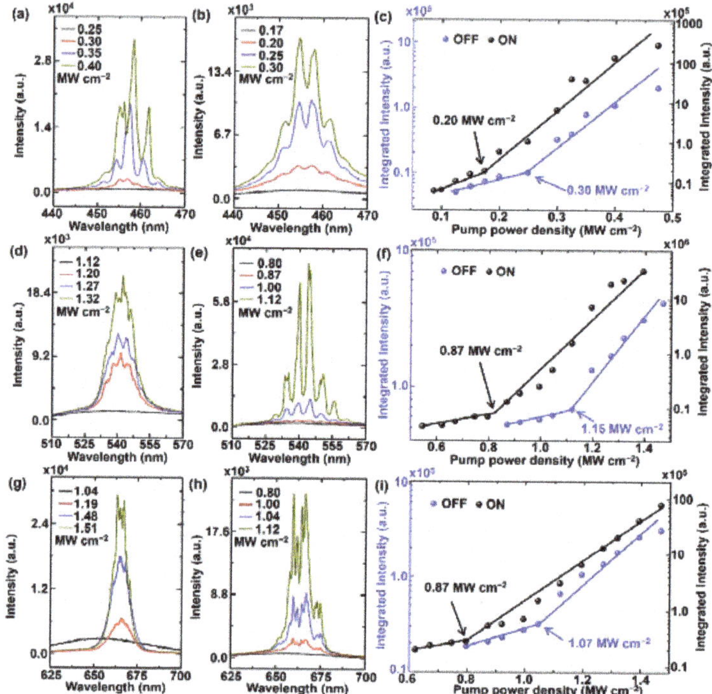

Figure 2. Performance characterization of the random lasers. (**a**,**b**) The emission spectra of RL_B at different pump power densities when the electric field is removed (**a**) and applied (**b**). (**c**) The integral intensity varies with pump power density for RL_B. (**d**,**e**) The emission spectra of RL_G at different pump power densities when the electric field is removed (**d**) and applied (**e**). (**f**) The integral intensity varies with pump power density for RL_G. (**g**,**h**) The emission spectra of RL_R at different pump power densities when the electric field is removed (**g**) and applied (**h**). (**i**) The integral intensity varies with pump power density for RL_R. The corresponding spectral integral range is from 440 nm to 470 nm for RL_B, from 510 nm to 570 nm for RL_G and from 625 nm to 700 nm for RL_R.

Similarly, the effect of scattering feedback of individual scattering part on the emission spectra of green random laser was also studied. A green random laser named as RL_G was obtained by coupling the C153 solution and PDLC film. When the electric field of PDLC film is removed (OFF), coherent feedback random lasing with a central wavelength of about 542.7 nm and a threshold of about 1.15 MW cm^{-2} can be observed (Figure 2d,f). Meanwhile, when the electric field is applied to the PDLC film (ON), coherent feedback random lasing with a threshold of about 0.87 MW cm^{-2} can be obtained (Figure 2e–f). The regulation effect of scattering feedback on emission spectra of RL_G is similar to that of RL_B. In addition, the effect of scattering feedback of individual scattering part on the emission spectra of red random laser is demonstrated. A red random laser named as RL_R was achieved by coupling the DCJTB solution and PDLC film. When the electric field of

PDLC film is removed (OFF), coherent feedback random lasing with a threshold of about 1.07 MW cm^{-2} can be observed around 662.9 nm in Figure 2g,i. When the PDLC film is applied electric field and is in ON state, coherent feedback random lasing with a threshold of about 0.87 MW cm^{-2} is obtained (Figure 2h,i). The experimental results show that the random lasing can always be achieved based on the separate coupling configuration with pure gain region and PDLC scattering part. Moreover, the threshold of the blue, green and red random lasing can be effectively modified through regulating the scattering feedback of individual scattering part.

The effect of scattering feedback of individual scattering part on the intensity of random lasing is demonstrated by continuously pumping 100 times at a fixed pump power density in Figure 3. The integral intensities of the emission spectra are calculated when the PDLC film is with external electric field (ON, black sphere) and without electric field (OFF, blue sphere) at a pump power density of 0.62 MW cm^{-2} in Figure 3a. The emission intensity of RL_B in ON state is much higher than that in OFF state, and the average integral intensity of 100 spectra in ON state is about 109 times of that in OFF state. Similarly, RL_G and RL_R are continuously pumped for 100 times under the pump power densities of 1.39 MW cm^{-2} and 1.46 MW cm^{-2}, respectively. Their integral intensities of the emission spectra are demonstrated through when the electric field is applied (ON, black sphere) and removed (OFF, blue sphere) in Figure 3b,c. The scattering feedback of individual scattering part has a similar regulation effect on the emission intensity of RL_G and RL_R as RL_B. The average integral intensities of RL_G and RL_R in ON state are about 79 and 48 times higher than those in OFF state, respectively. Therefore, the intensity of the random lasing can be dynamically manipulated by applying or removing an external electric field to the PDLC film.

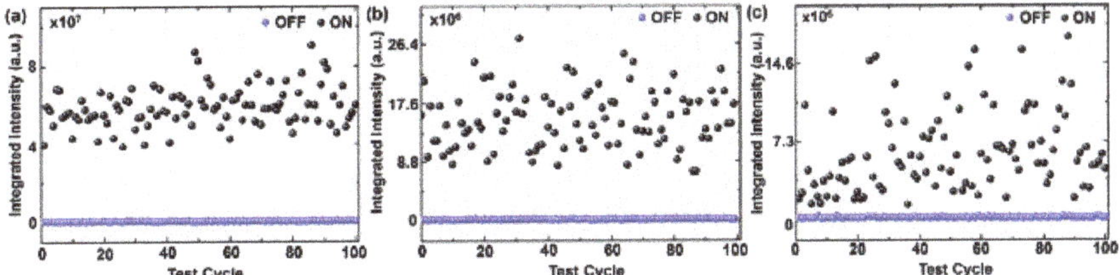

Figure 3. Regulation effect of scattering feedback on the intensity of random lasing. (**a**) The integral intensity evolution of the emission spectra when the RL_B is in ON state (black) or OFF state (blue) under a pump power density of 0.62 MW cm^{-2}. (**b**) The integral intensity evolution of the emission spectra when the RL_G is in ON state (black) or OFF state (blue) at 1.39 MW cm^{-2}. (**c**) The integral intensity evolution of the emission spectra when the RL_R is in ON state (black) or OFF state (blue) at 1.46 MW cm^{-2}. The corresponding integral range is (440, 470) nm for RL_B, (510, 570) nm for RL_G and (625, 700) nm for RL_R.

The effect of scattering feedback of individual scattering part on the spatial coherence of random lasing is also demonstrated by calculating the visibility of interference fringe. The Young's double-slit interference optical path is used to study the spatial coherence of the obtained random lasing. Two slits with a width of 50 μm are separated by 160 μm, and a CCD is positioned behind the slit plane to measure the far-field interference pattern. The visibility (V) of the interference fringe in the central region is calculated by $V = (I_{max} - I_{min})/(I_{max} + I_{min})$, where I_{max} and I_{min} are the local maximum and minimum values of the interference intensity near the observation point, respectively [21]. For example, the optical photo and the corresponding intensity distribution of the interference fringe generated by the blue random lasing at the pump power density of 0.35 MW cm^{-2} are shown in Figure 4a,b. The V is about 0.35 for the random lasing related to ON state, and about 0.16 for the random lasing under OFF state. The results show that the spatial coherence of blue random lasing in OFF state is much lower than that in ON state. Similarly, the optical photos and intensity distribution of the interference fringes obtained by the green random lasing passing through the double-slit interference optical path at the pump power density of 1.32 MW cm^{-2} are shown in Figure 4c,d. The V are about 0.34 (ON) and 0.16 (OFF), respectively. In addition, the spatial coherence of the red random lasing is also characterized at 1.32 MW cm^{-2}. The V of the interference fringes of red random lasing are 0.31 (ON, Figure 4e) and 0.14 (OFF, Figure 4f), respectively. The experimental results show that the visibility of interference fringes of the obtained random lasing in ON state is greater than that in OFF state. Moreover, the spatial coherence of random lasing emitting from another side are also measured, and the results prove that there exists the weak coupling between the gain region and the scattering region of the PDLC film supplying tunable backscattering. Therefore, the spatial coherence of the obtained random lasing can be dynamically manipulated by electrically regulating the scattering coefficient of the PDLC scattering part.

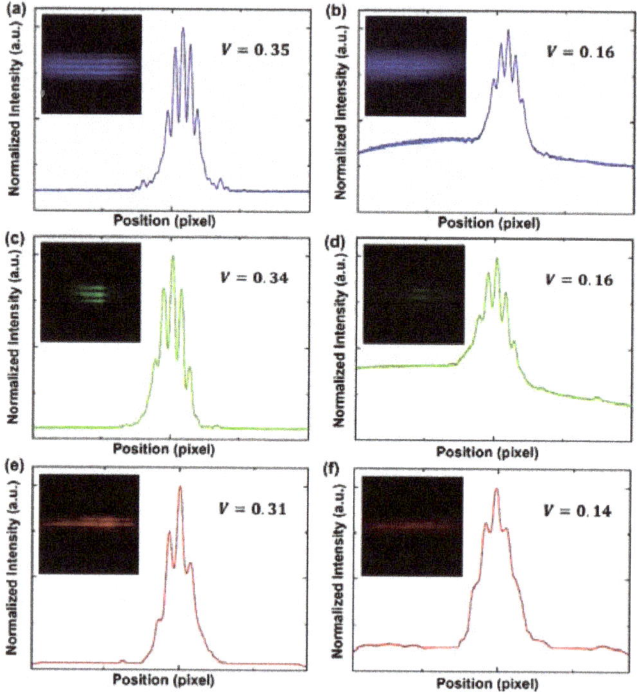

Figure 4. Regulation effect of scattering feedback on the spatial coherence of random lasing. (**a**,**b**) The intensity distribution and visibility of the interference fringes obtained by the double slit interference of RL_B under ON state (**a**) or OFF state (**b**). (**c**,**d**) The intensity distribution and visibility of the interference fringes for the RL_G under ON state (**c**) or OFF state (**d**). (**e**,**f**) The intensity distribution and visibility of the interference fringes for the RL_R under ON state (**e**) or OFF state (**f**). Illustrations are the optical photos of the interference fringes, respectively.

4. Conclusions

In summary, a random lasing with tunable spatial coherence is designed based on the separate coupling between gain medium and scattering part. The proposed random laser has the advantages of flexible controllability, easy fabrication and recyclability. First, when the PDLC film is separately coupled with different color gain dyes, random lasing with different wavelengths is realized under the excitation of pulsed laser. Then, the scattering intensity of PDLC film is manipulated by applying or removing electric field, and the threshold and intensity of random lasing are regulated. Finally, the spatial coherence of the random lasing can also be dynamically regulated. In short, the intensity and spatial coherence of random lasing can be manipulated by controlling the electric field. The proposed random lasing is expected to be used in interference measurement, security, optical imaging and display.

Author Contributions: Conceptualization, Z.W. and W.G.; Investigation, Y.B. and J.Z.; Methodology, Y.B. and D.L.; Project administration, Z.W.; Supervision, Z.W.; Validation, Z.W.; Visualization, Y.B. and Z.W.; Writing, Y.B., H.Y. and Z.W. All authors have read and agreed to the published version of the manuscript.

Funding: This research was funded by National Natural Science Foundation of China (grant Nos. 92150109, 11574033 and 61975018).

Acknowledgments: The authors acknowledge support from the National Natural Science Foundation of China and Beijing Higher Education Young Elite Teacher Project.

Conflicts of Interest: The authors declare no conflict of interest.

References

1. Bian, Y.; Shi, X.; Hu, M.; Wang, Z. A ring-shaped random laser in momentum space. *Nanoscale* **2020**, *12*, 3166–3173. [CrossRef] [PubMed]
2. Redding, B.; Choma, M.A.; Cao, H. Speckle-free laser imaging using random laser illumination. *Nat. Photonics* **2012**, *6*, 355–359. [CrossRef] [PubMed]
3. Lee, Y.-J.; Yeh, T.-W.; Yang, Z.-P.; Yao, Y.-C.; Chang, C.-Y.; Tsai, M.-T.; Sheu, J.-K. A curvature-tunable random laser. *Nanoscale* **2019**, *11*, 3534–3545. [CrossRef] [PubMed]
4. Mermillod-Blondin, A.; Mentzel, H.; Rosenfeld, A. Time-resolved microscopy with random lasers. *Opt. Lett.* **2013**, *38*, 4112–4115. [CrossRef]
5. Hou, Y.; Zhou, Z.H.; Zhang, C.H.; Tang, J.; Fan, Y.Q.; Xu, F.F.; Zhao, Y.S. Full-color flexible laser displays based on random laser arrays. *Sci. China Mater.* **2021**, *64*, 2805–2812. [CrossRef]
6. Xu, Z.; Hong, Q.; Ge, K.; Shi, X.; Wang, X.; Deng, J.; Zhou, Z.; Zhai, T. Random Lasing from Label-Free Living Cells for Rapid Cytometry of Apoptosis. *Nano Lett.* **2022**, *22*, 172–178. [CrossRef]
7. Hu, M.; Bian, Y.; Shen, H.; Wang, Z. A humidity-tailored film random laser. *Org. Electron.* **2020**, *86*, 105923. [CrossRef]
8. Shi, X.; Song, W.; Guo, D.; Tong, J.; Zhai, T. Selectively Visualizing the Hidden Modes in Random Lasers for Secure Communication. *Laser Photonics Rev.* **2021**, *15*, 2100295. [CrossRef]
9. Sznitko, L.; Chtouki, T.; Sahraoui, B.; Mysliwiec, J. Bichromatic Laser Dye As a Photonic Random Number Generator. *ACS Photonics* **2021**, *8*, 1630–1638. [CrossRef]
10. Boschetti, A.; Taschin, A.; Bartolini, P.; Tiwari, A.K.; Pattelli, L.; Torre, R.; Wiersma, D.S. Spectral super-resolution spectroscopy using a random laser. *Nat. Photonics* **2020**, *14*, 177–183. [CrossRef]
11. Turitsyn, S.K.; Babin, S.A.; Churkin, D.V.; Vatnik, I.D.; Nikulin, M.; Podivilov, E.V. Random distributed feedback fibre lasers. *Phys. Rep.* **2014**, *542*, 133–193. [CrossRef]
12. Cao, H.; Eliezer, Y. Harnessing disorder for photonic device applications. *Appl. Phys. Rev.* **2022**, *9*, 011309. [CrossRef]
13. Luan, F.; Gu, B.; Gomes, A.S.; Yong, K.-T.; Wen, S.; Prasad, P.N. Lasing in nanocomposite random media. *Nano Today* **2015**, *10*, 168–192. [CrossRef]
14. Ma, X.Y.; Ye, J.; Zhang, Y.; Xu, J.M.; Wu, J.; Yao, T.F.; Leng, J.Y.; Zhou, P. Vortex random fiber laser with controllable orbital angular momentum mode. *Photonics Res.* **2021**, *9*, 266–271. [CrossRef]
15. Tong, J.; Li, S.; Chen, C.; Fu, Y.; Cao, F.; Niu, L.; Zhai, T.; Zhang, X. Flexible Random Laser Using Silver Nanoflowers. *Polymers* **2019**, *11*, 619. [CrossRef]
16. Lee, K.; Ma, H.J.; Rotermund, F.; Kim, D.K.; Park, Y. Non-resonant power-efficient directional Nd:YAG ceramic laser using a scattering cavity. *Nat. Commun.* **2021**, *12*, 8. [CrossRef]
17. Wang, Y.C.; Li, H.; Hong, Y.H.; Hong, K.B.; Chen, F.C.; Hsu, C.H.; Lee, R.K.; Conti, C.; Kao, T.S.; Lu, T.C. Flexible Organometal-Halide Perovskite Lasers for Speckle Reduction in Imaging Projection. *ACS Nano* **2019**, *13*, 5421–5429. [CrossRef]
18. Liu, Y.L.; Yang, W.H.; Xiao, S.M.; Zhang, N.; Fan, Y.; Qu, G.; Song, Q.H. Surface-Emitting Perovskite Random Lasers for Speckle-Free Imaging. *ACS Nano* **2019**, *13*, 10653–10661. [CrossRef]
19. Redding, B.; Choma, M.A.; Cao, H. Spatial coherence of random laser emission. *Opt. Lett.* **2011**, *36*, 3404–3406. [CrossRef]
20. Ghofraniha, N.; La Volpe, L.; Van Opdenbosch, D.; Zollfrank, C.; Conti, C. Biomimetic Random Lasers with Tunable Spatial and Temporal Coherence. *Adv. Opt. Mater.* **2016**, *4*, 1998–2003. [CrossRef]
21. Ismail, W.Z.W.; Liu, D.M.; Clement, S.; Coutts, D.W.; Goldys, E.M.; Dawes, J.M. Spectral and coherence signatures of threshold in random lasers. *J. Opt.* **2014**, *16*, 105008. [CrossRef]
22. Bian, Y.X.; Xue, H.Y.; Wang, Z.N. Programmable Random Lasing Pluses Based on Waveguide-Assisted Random Scattering Feedback. *Laser Photonics Rev.* **2021**, *15*, 2000506. [CrossRef]

23. Gottardo, S.; Cavalieri, S.; Yaroshchuk, O.; Wiersma, D.S. Quasi-two-dimensional diffusive random laser action. *Phys. Rev. Lett.* **2004**, *93*, 263901. [CrossRef] [PubMed]
24. Xiao, S.M.; Song, Q.H.; Wang, F.; Liu, L.Y.; Liu, J.H.; Xu, L. Switchable random laser from dye-doped polymer dispersed liquid crystal waveguides. *IEEE J. Quantum Electron.* **2007**, *43*, 407–410. [CrossRef]
25. Wiersma, D.S.; Cavalieri, S. Light emission: A temperature-tunable random laser. *Nature* **2001**, *414*, 708–709. [CrossRef]
26. Chen, S.J.; Shi, J.W.; Kong, X.Y.; Wang, Z.N.; Liu, D.H. Cavity coupling in a random laser formed by ZnO nanoparticles with gain materials. *Laser Phys. Lett.* **2013**, *10*, 55006. [CrossRef]
27. Shi, X.Y.; Bian, Y.X.; Tong, J.H.; Liu, D.H.; Zhou, J.; Wang, Z.N. Chromaticity-tunable white random lasing based on a microfluidic channel. *Opt. Express* **2020**, *28*, 13576–13585. [CrossRef]
28. Shi, X.Y.; Chang, Q.; Bian, Y.X.; Cui, H.B.; Wang, Z.N. Line Width-Tunable Random Laser Based on Manipulating Plasmonic Scattering. *ACS Photonics* **2019**, *6*, 2245–2251. [CrossRef]

MDPI
St. Alban-Anlage 66
4052 Basel
Switzerland
Tel. +41 61 683 77 34
Fax +41 61 302 89 18
www.mdpi.com

Crystals Editorial Office
E-mail: crystals@mdpi.com
www.mdpi.com/journal/crystals

www.ingramcontent.com/pod-product-compliance
Lightning Source LLC
LaVergne TN
LVHW070632170726
843515LV00004B/235

* 9 7 8 3 0 3 6 5 7 6 7 2 5 *